T0211264

Building Embedded Systems

Programmable Hardware

Changyi Gu

Apress®

Building Embedded Systems

Changyi Gu
San Diego
California, USA

ISBN-13 (pbk): 978-1-4842-1918-8   ISBN-13 (electronic): 978-1-4842-1919-5
DOI 10.1007/978-1-4842-1919-5

Library of Congress Control Number: 2016941346

Managing Director: Welmoed Spahr
Lead Editor: Jonathan Gennick
Development Editor: Douglas Pundick
Technical Reviewer: Brendan Horan
Editorial Board: Steve Anglin, Pramila Balen, Louise Corrigan, Jim DeWolf, Jonathan Gennick,
    Robert Hutchinson, Celestin Suresh John, Michelle Lowman, James Markham, Susan McDermott,
    Matthew Moodie, Jeffrey Pepper, Douglas Pundick, Ben Renow-Clarke, Gwenan Spearing
Coordinating Editor: Jill Balzano
Copy Editor: Kezia Endsley
Compositor: SPi Global
Indexer: SPi Global
Artist: SPi Global

Distributed to the book trade worldwide by Springer Science+Business Media New York, 233 Spring Street, 6th Floor, New York, NY 10013. Phone 1-800-SPRINGER, fax (201) 348-4505, e-mail orders-ny@springer-sbm.com, or visit www.springer.com. Apress Media, LLC is a California LLC and the sole member (owner) is Springer Science + Business Media Finance Inc (SSBM Finance Inc). SSBM Finance Inc is a Delaware corporation.

For information on translations, please e-mail rights@apress.com, or visit www.apress.com.

Apress and friends of ED books may be purchased in bulk for academic, corporate, or promotional use. eBook versions and licenses are also available for most titles. For more information, reference our Special Bulk Sales–eBook Licensing web page at www.apress.com/bulk-sales.

Any source code or other supplementary material referenced by the author in this text is available to readers at www.apress.com. For detailed information about how to locate your book's source code, go to www.apress.com/source-code/.

Printed on acid-free paper

To those who enjoy doing things that are otherwise meaningless in the eyes of many.

Contents at a Glance

Contents

About the Author

Changyi Gu has worked for multiple high-tech companies across California for over 15 years. He is also the founder of PulseRain Technology, LLC, a company that excels at FPGA-based embedded systems. He started out his career in firmware development. With a deep passion in engineering, he later devoted himself to wireless hardware and programmable logic. Thus, he is one of the lucky few who gets to watch things closely on both the software and hardware sides.

Changyi graduated from the University of Southern California with a master's degree in VLSI. And he holds another master's degree in Communications and a bachelor's degree in CS from Shanghai Jiaotong University, China. In his leisure time, he likes to tinker with things.

About the Technical Reviewer

 **Brendan Horan** is a hardware fanatic, with a full high rack of all types of machine architectures in his home. He has more than 10 years of experience working with large UNIX systems and tuning the underlying hardware for optimal performance and stability. Brendan's love for all forms of hardware has helped him throughout his IT career, from fixing laptops to tuning servers and their hardware in order to suit the needs of high-availability designs and ultra low-latency applications. Brendan takes pride in the Open Source Movement and is happy to say that every computer in his house is powered by open source technology. He resides in Hong Kong with his wife, Vikki, who continues daily to teach him more Cantonese.

Acknowledgments

Family:

First of all, I would like to thank my parents for nurturing my interest in engineering since childhood, which has given rise to this book.

I would also like to thank my wife and daughter for bearing with my nocturnal lifestyle for so long. The midnight oil I burned is in your names.

Tech Reviewers:

Special thanks to Brendan Horan for his sound suggestions to my writing. Those feedbacks are highly valued.

And I also owe a big thank you to my ex-coworkers—Cynthia Xu and On Wa Yeung—for offering assistance in reading my book. I really appreciate your kind help.

Editors:

A hug to Jonathan Gennick for giving me the opportunity on this book, and for his passion in embedded systems. I really appreciate your prompt response to my book proposal and your hard work throughout the whole project.

I would also like to thank Jill Balzano for her wonderful job managing the project. This book will not be possible without you.

My development editor, Douglas Pundick, offered valuable suggestions to save me from possible embarrassment. Your thoroughness is appreciated.

And to the rest of Apress team—thanks for your hard work!

Readers:

Thank you for choosing my book!

Introduction

"Almost no one comes down here (the engineering level) unless, of course, there is a problem. That's how it is with people. Nobody cares how it works as long as it works. I like it down here."

—The Matrix Reloaded (2003)

Embedded systems are conspicuous by their omnipresence. But most people take them for granted despite the fact that behind every system is a design that involves multiple engineering disciplines. Even something as humble as a toaster might be the joint effort of software and hardware engineers. Assuming you still have the unabated curiosity to know how stuff works, this book will try to fill any knowledge gaps you might run into.

Actually, this book is a brain dump of what I have picked up over the past 15+ years. Some of this knowledge has been learned the hard way, which is passively through insomnia and hair loss. And it is my best wishes that you don't have to go through the same ordeals I endured.

In addition to the traditional design approach, the semiconductor industry's latest juggernaut has made SOPC (System on Programmable Chips) a viable option both technologically and financially. It is a trail blazed by many newly designed embedded systems, which this book explores in depth.

Lastly, this book will serve you better if you've taken some college-level courses in CS or EE, or maybe a little bit of both. Basic knowledge in C/C++ and FPGA can help you sail through this book more smoothly.

Thank you for your faith in my book!

—Changyi Gu
02/14/2016 in San Diego, CA

CHAPTER 1

The Whole Picture

A complex system that works is invariably found to have evolved from a simple system that worked.

—John Gall (Ref [1])

The world of *embedded systems* is pretty much a farrago of everything, with each system having its own idiosyncrasies. In the old days, many systems would forgo microprocessors in favor of glue logic or application specific controllers to drive costs down. This makes it very hard to have a generalized discussion about those systems.

However, things have changed since then. The past 20 years witnessed the relentless push of semiconductor processes. In the mid-1990s, 0.35μm (350nm) was considered the state of the art. And when this book was being conceived, 14nm FinFET had just begun mass production. Big Fabs like Intel, TSMC, and Samsung are elbowing their way toward 10nm. The juggernaut of semiconductor technology has greatly shaped the landscape of embedded systems in the following ways:

- The microprocessor's clock rate has increased exponentially while its power consumption dropped. Thanks to its high production volume, unit price has also plummeted. Thus microprocessors have managed to gain the universal presence in today's embedded systems.

- However, the sophistication of the semiconductor process has also boosted the NRE (Non-Recurring Engineering) costs dramatically. And the high NRE costs have to be balanced out by high volumes in order to make the ends meet. This makes the chip tape-out riskier and less affordable. Consequently, for products with small- to mid-range volumes, programmable logic devices, such as FPGA, become more and more favorable over ASIC.

- For the semiconductor process, shrinking feature size means higher density, which makes it possible for processors and programmable logic to be integrated on the same die. Major FPGA vendors now all have programmable SoC (System on Chip) devices in their portfolios[1] to target various market segments. And when this book was being conceived, Intel has just acquired Altera for more than $16 billion to "improve chip performance" with programmable logic devices.

Electronic supplementary material The online version of this chapter (doi:10.1007/978-1-4842-1919-5_1) contains supplementary material, which is available to authorized users.

---

[1]Altera offers HPS (Hard Processor System) in almost all of its high end products. Xilinx has Zynq serial SoC FPGA. Microsemi provides SmartFusion serial SoC FPGA.

© Changyi Gu 2016
C. Gu, *Building Embedded Systems*, DOI 10.1007/978-1-4842-1919-5_1

As a result, the traditional boundary between software and hardware has started to blur. For embedded systems practitioners, cross-disciplinary knowledge and skills are increasingly valued. Having that in mind, this book will try to cover both grounds with a practical approach.

As stated previously, thanks to the ferocious push of semiconductor technology, it's no longer the luxury of high-end design to have a microprocessor on board. The prevailing truth is that nowadays the majority of embedded systems have at least one microprocessor on board, and they share many common traits in this regard. Assuming the system under discussion is a uni-processor system, its general architecture can be aptly illustrated in Figure 1-1.

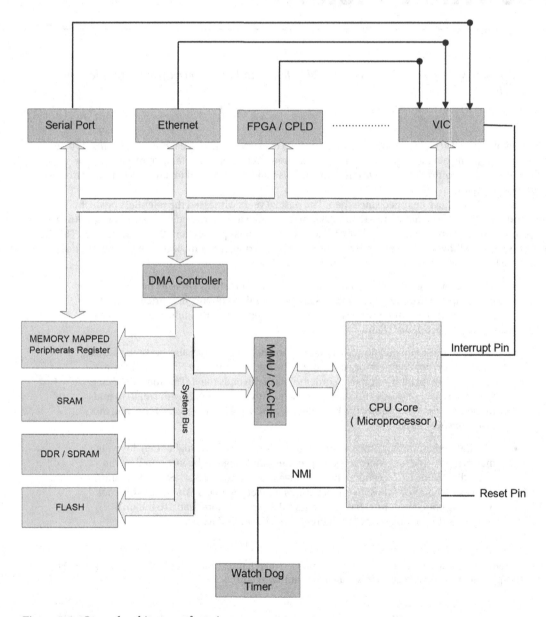

Figure 1-1. *General architecture of a uni-processor system*

A few notes about the image in Figure 1-1:

- Keep in mind that due to the high density of today's semiconductor process, the majority of the components shown in Figure 1-1 could be integrated in the same package instead of being multiple discrete components on board.

- Many technical details are omitted from Figure 1-1 in order not to distract readers from the whole picture. Those technical details are covered in later chapters when the correspondent modules are mentioned.

- The system bus in Figure 1-1 could be AMBA (if it is a SoC design with ARM CPU core), PCI, PCI-express, or any other bus architecture.

- Figure 1-1 also assumes there is a VIC (Vector Interrupt Controller) in the system, although this is not always the case. The topic of interrupt controllers is covered in later chapters.

- As mentioned, programmable logic has started to gain traction as the NRE cost of ASIC has skyrocketed. Although not every system has FPGA or CPLD, FPGA/CPLD devices will be a valuable asset to extend system functions. In fact, today's FPGA is so resourceful that it could either absorb the microprocessor into its programming logic as a soft core (Ref [2]) or integrate hardcore processors with the programming logic in one package (Ref [3][4]). On top of that, some FPGA vendors go the extra mile by providing analog frontend in their products (Ref [4]). Given all that, a great amount of ink will be spilled on FPGA/CPLD later in this book.

- This book does not explicitly distinguish among "microprocessor," "processor," "CPU" and "CPU core," since their differences are insignificant for all practical purposes. Thus the rest of the book will use those terms interchangeably. However, this book does make a distinction between the term "microprocessor" and "microcontroller," as the former only refers to the processor core, while the latter usually means the processor core plus RAM/ROM and other peripherals in a single package.

Organization of the Book

The rest of the book will be organized as follows:

- Chapters 2–5 will illustrate how things work behind the scenes, starting from the moment when the power switch is flipped on.

- Afterward, two chapters will discuss embedded software (firmware) programming, with one chapter covering C language and the other covering C++. There is also a chapter on firmware build and deployment.

- SOPC/FPGA is an import topic of this book. Chapter 9 will start on FPGA development, followed by another chapter on the SOPC design approach.

- Other miscellaneous hardware topics, such as power management and LCD display, are discussed in Chapter 11.

- Math is always the necessary evil—please excuse me for the lack of better words—for engineers. A whole chapter will be devoted to fixed-point math.

- What sets human beings apart from other species is that we know how to use tools. Thus a separate chapter covers tool preparation.

- Consistency and reliability are highly valued in any engineering effort. A discussion on workflow helps to conclude this book.

Companion Materials for this Book

Companion materials, such as sample code, supporting scripts, etc., will be released as open source material. You may get them on the Apress web site (`http://www.apress.com`) or through `http://open.pulserain.com` (link to GitHub).

References

1. *Systematics—How Systems Work and Especially How They Fail,* by John Gall, Quadrangle/The New York Times Book Co., 1977
2. *Nios II Gen2 Processor Reference Guide,* Altera Corporation, April, 2015
3. *Zynq-7000 All Programmable SoC: Embedded Design Tutorial,* UG1165 (v2015.1), Xilinx Inc. April, 2015
4. *SmartFusion Customizable System-on-Chip (cSoC) Rev 13,* Microsemi Corporation, March 2015

CHAPTER 2

Power On and Bootloader

The universe exists because of spontaneous creation.

—Stephen Hawking

Curiosity is human nature. In a grand scheme of things, people might wonder how the universe got started, to which those folks who discovered the gravity wave know better than to ask. But on a much smaller scale, if you are just curious to understand how the embedded systems got started, this chapter can help.

What Happens After Powering On

The first time you get your hands on a brand new smartphone, you push the power button, see the screen illuminated, and then see a fancy welcome page or logo dazzling in front your eyes. Have you ever wondered what's going on from the moment you turn on the power to the minute when you see the fancy GUI? Who or what is behind all this?

For the majority of modern day embedded systems, there is at least one processor on board. When the power is turned on, the processor is getting hard-reset. The processor's program counter is reset to a predetermined value. In other words, the processor starts to fetch instructions from a fixed address (called the *reset handler*), and it starts to execute that reset handler after power on.

Take ARM9 (32-bit CPU) for example (Ref [1]). Its PC (Program Counter) would be reset to either 0x00000000 or 0xFFFF0000, depending on the hardware pin configuration. The space allocated for its reset handler is 4 bytes, which is exactly the size of one instruction. A branch instruction or a load PC (LDR PC) instruction is the usual candidate. It points the program counter to the entry of another bigger program, called the *bootloader. That is to say, the first program to be executed after power on reset is the bootloader.*

Depending on the actual hardware pin configuration, the mapping of the reset handler address can be completed in one of the ways:

1. The address is mapped to some on-chip ROM, with factory programmed instructions.

2. The address is mapped to off-chip memory. Usually the off-chip memory is some sort of non-volatile memory, like Flash or EEPROM. Take ARM9, for example—if its reset handler is configured to 0x00000000, a Flash memory that contains bootloader can be mapped to that address (see Figure 2-1).

© Changyi Gu 2016
C. Gu, *Building Embedded Systems*, DOI 10.1007/978-1-4842-1919-5_2

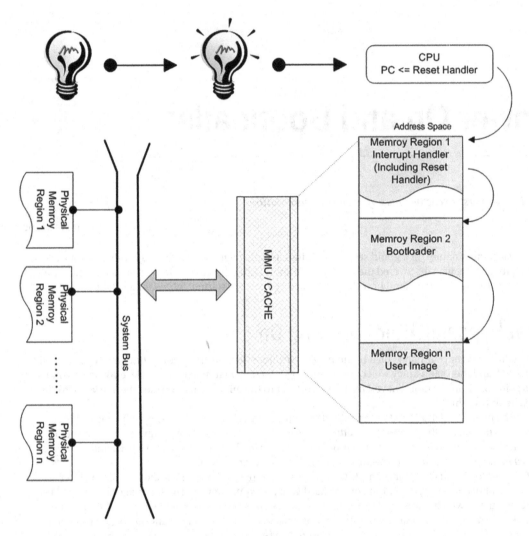

Figure 2-1. *What happens after a power on reset*

3. No matter where the reset handler is mapped, *the users can always use ICE (In-Circuit Emulator) as the bootloader.* Usually ICE loads the image through a JTAG interface. (ICE will be discussed in more detail in later chapters.)

The bottom line is that bootloader is the first program to be executed after a power on reset.

What Does Bootloader Do?

Bootloader's tasks makes for a long list, including but not limited to the following:

1. *Initialize the memory.*

   If your system has memory types like DDR, chances are the DDR needs to be initialized before it can have the first r/w access. Take personal computers for example. Some PCs will print out detailed hardware information at boot time. The typical message may look like `"DRAM Clock=xxxMHz, SDRAM CAS Latency=y.z"`. That means the BIOS (the PC's bootloader) has configured your DDR with these parameters. (On a PC, it is the DIMM you plug into the memory slot.)

   Also keep in mind that ICE will not initialize the DDR automatically for you. So if you want to load an image to DDR through ICE, make sure DDR is properly configured.

2. *Initialize the user terminal or debug console.*

   Your system has to find a way to output something, which can be as conspicuous as a high-resolution screen, or as humble as an RS232 console. Although many systems do not have screen output, they still require serial console to output debug messages or get input at the development stage. Bootloader is responsible for setting them up.

3. *Initialize other peripherals.*

   Depending on the nature of your system, other peripherals such as Ethernet also need to be configured. Sometimes bootloader can load the user image through Ethernet or RS232 instead of using Flash memory.

4. *Configure cache and MMU.*

   As you can see in Figure 2-1, memory access has to go through cache and MMU. As far as bootloader is concerned, performance is not its first priority. Thus bootloaders could choose to disable the cache and MMU for the peace of mind. After the user image is loaded and executed, cache and MMU are re-enabled in order to gain better performance.

5. *Provide basic services to the users.*

   Most commercial bootloaders are able to provide some basic commands to end users, which allow users to configure the Flash or check memory content. Because of these added capabilities, developers sometimes call these commercial bootloaders Debug Monitor or ROM Monitor.

6. *Provide service calls to OS.*

   Not every bootloader does this. In fact, most embedded bootloaders are decoupled with the embedded OS. However, on the IBM PC platform, the BIOS would actually provide some basic interrupt calls to OS. A typical case for this is the hard disk access. When you are booting from your hard disk to load Windows, you might have to wait a short while before you can see the Windows logo. In fact, BIOS is doing the disk r/w during that period and the Windows device driver takes over by the time when you see the logo. BIOS might do the disk access in PIO mode since bootloaders usually prefer simplicity over performance, but the Windows device driver could do every r/w in DMA mode to get the highest throughput.

7. *Load the user image into memory.*

As the name "bootloader" suggests, *the main function of a bootloader is to load the user image into memory and execute it from the entry address.* The user image could be as big as a full-blown embedded OS, or as small as a tiny test program, and it can be loaded from Flash, Ethernet, a serial port, or any other available interface.

User Image (ROM Image)

As it was mentioned at the end of the last section, bootloader can load the user image into memory. However, be advised that "user image" is a generic term. There are a few cases that have to be distinguished from each other when it comes to the details.

Image of Embedded OS

If the image to be loaded is an image of the embedded OS (such as the "zImage" of Linux), bootloader treats it as raw binary. That is to say, bootloader will load the image as a big trunk of binary and copy it to the destination address blindly.

For example, to load Linux zImage under redboot (Ref [4]), the command will be something like:

```
Redboot> load -r -b 0x01234000 zImage
Redboot> go 0x01234000
```

Here the -r option means it is raw binary and the -b option provides the destination address. Then the go command starts to execute from the entry address.

For an embedded OS, its image is built with init code positioned at the beginning part of the image. Most likely it will be compressed as well to keep a small footprint. The init code would do some preliminary setting and then decompress the rest of the image. It would then extract and load OS kernel into memory, as illustrated in Figure 2-2.

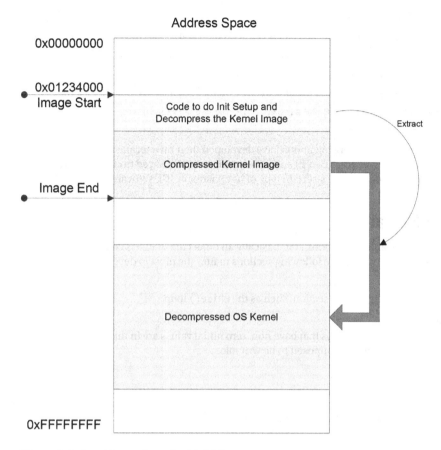

Figure 2-2. *Load image for embedded OS*

Image of Object File

If the image to be loaded is an object file produced by a cross-compiler, the image must be structured and conform to the standard format. Bootloader would no longer treat it as raw image. If the OS has already booted up when the object file is being loaded, the OS loader will replace the bootloader to load these object file images.

a.out, COFF, and ELF

The formats for object files have an evolutionary history of half a century. Among all the formats that have been developed, three carry the most weight: a.out, COFF, and ELF.

a.out is probably the oldest object file format on the UNIX system. Although it is no longer used today, as a legacy, many C compilers still use a.out as the default output filename.

COFF is the successor to a.out. Many of Texas Instruments' DSP processors use COFF as the default object file format (Ref [6]). However, COFF has imposed some limitations on the object file, and it is succeeded by the ELF format in large part.

ELF is now the default object file format for many cross-compiler tools, not to mention it is also the default format used on Linux.

■ **Labs** As mentioned earlier, many C compilers still use a.out as the default output filename, but the actual object file format might be something else, such as ELF. If you have a C file called test.c and you type gcc test.c on a UNIX command line, you will get a new file named a.out. If you type objdump --file-headers a.out afterward, it probably will tell you that the file named a.out is actually in a format other than a.out. (On the Sun server I'm using, it shows something like a.out: file format elf32-sparc.)

Over the years, many companies and organizations have developed their own formats based on these three, like Microsoft's PE format (derived from COFF) and ARM's AXF format (derived from ELF). Often times, these formats can only be recognized by a certain type of bootloader or ICE software.

Sections of the ELF Format

The prevalence of ELF format warrants a close look at it. Basically an object file in ELF is composed of many sections (Ref [7]). For all practical purposes, the following sections matter the most to developers:

- .text section
  All the executable code lies in this section, such as the while() loop.

- .data section
  All the global variables/static arrays that have non-zero initial values are in this section. Note that this section is supposed to be writable.

  Examples:

  ```
  int gReadFlag = 1;                    // Global variable
  int gBuffer[] = {1, 1, 2, 3, 5, 8};  // Global array
  ```

  Static variables/arrays with non-zero initial values are also located in this section, such as:

  ```
  int test()
  {
    static int start_value = 1;    // static variable
    static int abc[] = {1, 2, 3};  // static array

      ...

  } // End of test()
  ```

- .rodata section
  All the read-only data is placed in this section, such as global constant array. On DSPs, such as ARM9E, the read-only data should be loaded into I-TCM (Instruction Tightly Coupled Memory). As its name suggests, TCM lies close to the processor core. Hence, it has a very short access time. These are not supposed to change, just like the .text section.

  For example:

  ```
  const int gIndex[] = {1, 3, 7, 9}; // Global const array
  ```

- .bss section

  All the global/static variables that are not explicitly initialized in the code default to zero. The .bss section hosts these variables. When bootloader loads the object file, it will fill the .bss section with zeros.

  For example:

  ```
  int gReadFlag;       // Global variable, uninitialized
  int gReadBuffer[16]; // Global array, uninitialized
  ```

  Since this section is supposed to be all zeros initially, it does not have to take up physical space when you build a Flash image out of the object file.

Link Script

When bootloader loads an object file, it is actually loading these aforementioned sections from the object file into the RAM. So the destination address and size of the destination memory block have to be specified for each section. What does the target system's address space look like? How do you fill the memory in the target system with these sections? When you are building the object file, these details are all stipulated in a text file called *link script file* or *link editor file,* which is used by the linker. The linker will use its default link script if one is not provided. (Under ARM's RVDS, this link script is called a *scatter file*, with .scat filename extension.)

With GNU linker, you could see the default link script by typing:

```
ld --verbose
```

There are two things that have to be done in a link script— you must define the memory in the target system and you must determine the address of each section in the object file.

Defining the Memory in the Target System

The link script has to tell the linker how many physical memory segments exist in the target system, as well as the size and starting address of each segment. Linker does not care about the physical attributes of these memory segments. In other words, no-volatile memory and volatile memory look the same in the eyes of linker. However, these physical attributes do matter to developers, especially when they are running code straight from no-volatile memory (XIP, which would be elaborated in later chapters) or the target memory needs to be initialized before the first access (like most DDR memory).

Listing 2-1 shows a snippet of GNU link script that defines the memory in the target systems.

Listing 2-1. MEMORY Command in GNU Link Script

```
MEMORY
{
    ROM  : org = 0x200,  len = 0x1000
    SRAM : org = 0x1300, len = 0x300
    DDR  : org = 0x1900, len = 0x200
}
```

In Listing 2-1, three memory segments were defined. Each segment should correspond to the actual memory map in the target system.

Determining the Address of Each Section in the Object File

The bootloader needs to know where to load each section in an object file. Such information has to be stored in the object file at link time.

Listing 2-2 shows a snippet of the GNU link script for this purpose.

Listing 2-2. SECTIONS Command in GNU Link Script

```
SECTIONS
{
  my_code :
  {*(.text);} >SRAM

  my_data :
    {*(.data);*(.rodata);} >SRAM

  .bss :
    {*(.bss);} >DDR

  other :
    {*(.*);} >SRAM
}
```

In Listing 2-2, three output sections are defined. The my_code output section collects all the .text sections and loads them into SRAM. The my_data output section collects .data and .rodata sections and loads them into SRAM as well. The .bss section would be in DDR. (Here, the memory segment of ROM, SRAM, and DDR are all defined in Listing 2-1.)

■ **Labs** Let's merge Listings 2-1 and 2-2 into one file and call it link_script01.x. This is the link script you are going to use for this lab. You'll use the following test program as a guinea pig. The same program will be used in later chapters for Flash image generation. Call this program lost.c, as shown in Listing 2-3.

Listing 2-3. Test Program: lost.c

```
unsigned char lost_candidates[] = {0x04, 0x08, 0x0F, 0x10, 0x17, 0x2A};
unsigned char jungle[128];
const int John_Locke = 0x7A8B9CAA;
int main(void)
{
     int test_signature = 0xAABBCCDD;
     lost_candidates [John_Locke] = 0;
} // End of main()
```

LAB 2-1: BUILDING AN ELF OBJECT FILE

You'll compile this program with the following command. (I tested it out on a virtual machine with 32-bit Linux installed, because most embedded processors have a word length less than 64 bits. If you run the lab on 64-bit Linux, it should still work except that some offset numbers might be slightly different from what's shown here. However, Cygwin may not work well for this lab because the default file format in Windows is PE format instead of ELF.)

```
$> gcc -fno-exceptions -c lost.c
$>
$> objcopy --remove-section=.comment lost.o
$>
$> ld -o my_output01.elf -T link_script01.x lost.o
$>
$> objdump -h my_output01.elf
```

Among those commands:

1. The first command will compile the code and remove the .eh_frame section that otherwise might be generated.

2. The second command will remove the .comment section from the object file. The .comment section contains information like the GCC version number, platform supported, etc., which serves no purpose other than bloating the image size. So it should be removed from the object file before link.

3/4. The third command will link the object file with your own link script. Then the fourth command will dump the headers for each section into my_output01.elf.

Table 2-1 shows the output of objdump[1].

Table 2-1. *Objdump Output with Link Script 01*

Idx	Name	Size	VMA	LMA	File off	Algn
0	my_code	0000001b	00001300	00001300	00000300	2**2
		CONTENTS, ALLOC, LOAD, READONLY, CODE				
1	my_data	0000000c	0000131c	0000131c	00000334	2**2
		CONTENTS, ALLOC, LOAD, DATA				
2	.bss	00000080	00001900	00001900	00000900	2**5
		ALLOC				
3	other	00000000	00001328	00001328	00000328	2**2
		CONTENTS, ALLOC, LOAD, CODE				

---

[1]Depending on the processor type and compiler version, the actual output might be slightly different from what is demonstrated here.

For now, let's put aside the difference between VMA and LMA and consider them as the load address for each section. As you can see, my_code and my_data are both in the SRAM segment starting at 0x1300. And .bss is in the DDR starting at 0x1900. If you tell ICE to load this image, these are the addresses that ICE will use.

Flash Image and XIP (Execute In Place)

A few subtle details are intentionally omitted from the previous example, and these details will be put under the microscope here. Before we get started, I want to raise two questions that concern many:

- On the first day when we were learning C language, we were told that main() function is the start of our program, which is largely true. Then we put a printf ("Hello World!") in the main() function to hail the world. For anyone who tries to learn C, such "Hello World" practice is probably the universal starting point. But have you ever wondered what happens before the main() function? Does the CPU just magically jump onto the main()? Who is setting up the calling stack? And where does the printf() come from?

- Now you know that bootloader is the first program to run after reset, so nobody is going to load the bootloader. Bootloader has to help itself. However, since it is the first program to run, bootloader is always stored in some type of non-volatile memory, such as NOR-Flash or EEPROM. How do you run code straight from read-only memory? (It is usually impractical to write a single word to Flash or EEPROM on the fly.) In other words, how do you handle the .data and .bss sections in this case?

C Runtime Library and C Standard Library

This issue is not only confined to programs running out of Flash. In fact, it is a broad question that any developer has to ask. However, if you are just developing applications being loaded by OS, the OS and the compiler will answer this question for you without your involvement. But for those more primitive programs that run from Flash after a power reset, the developer has to do more than just keep his eyes on the main() function.

Here's what happened (a line borrowed from Tony Shalhoub (Ref [8])):

For a program like the "Hello World" in Listing 2-4, the compiler has to find two things in addition to main() in order to pass the link stage: the C *runtime library* and the C *standard library*. (Sometimes "C standard library" is shortened to "C library".)

Listing 2-4. "Hello World"

```
#include "stdio.h"
void main()
{
    printf ("Hello World\n");

} // End of main()
```

The C runtime library is usually written in assembly language and takes the name like `crt0.s` or `startup.s`. As its name implies, it needs to set up the runtime environment before the code in the `main()` function can be executed, which includes:

1. Initializing the hardware, such as CPU cache and MMU, if necessary.

2. Setting up the stack.

3. Initializing the `.bss` section to all zeros in the memory.

4. Calling the `main()` function.

For application developers, the compiler would provide a default runtime library during the link stage, which is good enough in most cases. However, for programs like the bootloader, the developer has to provide its own runtime library in order to set up the environment correctly. It will be the first code to run after a power reset.

Now let's talk a little bit more about the `printf ("Hello World!")` in Listing 2-4. As some of you might have known, `printf()` is a routine defined in the ISO C standard. The implementation of those standard routines is defined in the C standard library, or C Library for short. If you use any of such standard routines in your program, compiler would then link the C standard library.

For the program in Listing 2-4, if this program is to be compiled and run as an application under an OS environment, the OS would initialize the console hardware long before this program is loaded. And the compiler would probably use its default C standard library at link time.

However, if this program is to be compiled and run out of Flash after power reset, like the bootloader is, there are a few things you have to be aware of:

- The C runtime library would need to initialize the console before the `main()` function is called in order for `printf()` to work properly. You have to provide your own runtime library to the compiler.

- You have to pay attention to the size and efficiency of those routines in the C standard library. For many embedded systems, there are very restrictive constraints on space and timing. Sometime, those routines in the default C standard library turn out to be too bulky or too slow. Or even worse, they could have hidden surprises in store.

In order to reduce the footprint or save CPU cycles (reduce MIPS), you can do the implementation yourself instead of relying on the standard library. Or if you have to, you could use a different C standard library that is optimized for embedded development.

For example, there are math functions provided by standard library, such as `sin()` or `sqrt()`. However, some standard library may implement them in a slow fashion, such as by using an iteration method. In order to meet real-time requirements, you could rewrite these math functions using a look-up table, which would be a lot faster.

Also, there are some C standard libraries that are optimized for embedded systems. One of these libraries is µClibc, which has a smaller footprint than the GNU C Library (Ref [9]).

Watch out for `printf()`! Depending on the particular flavor of implementation, `printf()` could sneak a lot of fat into your image size. And it might also demand a big stack size (Ref [25]).

The bulkiness of `printf()` is mainly caused by the need to support a wide variety of format options (Ref [26]). If size is a concern, you could use a C library that implements a simplified version of `printf()`. Or, if applicable, you could choose not to use `printf()` at all and use something like `puts()` instead.

VMA and LMA

If your image is intended to be loaded into RAM and then executed, you only need to tell the linker one address. In other words, the RAM address you load your image into is also the address you execute your code from. There is no ambiguity here. You can store your image in any type of non-volatile storage, be it a Flash or a hard disk. (Note that in this case, if you store your image on a Flash drive, the Flash does not have to be mapped into the address space as long as you have a way to access it. You can even have file system on the Flash.)

However, things are more complicated if you execute code straight from non-volatile memory, such as Flash or ROM. Under such circumstances, the non-volatile memory must be mapped into the system address space. (This is called XIP—Execute In Place.)

Now this begs the question: For any XIP image, how do you deal with writable sections, such as .data and .bss?

When you are running code from Flash or ROM, you have to copy the .data section from Flash or ROM into RAM. Also, you have to initialize the .bss to all zeros in RAM. Such copy and initialization are supposed to be done by the runtime library or any startup code. (crt0.s or startup.s, as mentioned). Because of such copying, you have to provide two addresses in the link script for the .data sections. One is the ROM address where it is stored, which we call *LMA* (Load Memory Address). The other is the RAM address where it is accessed from, which we call *VMA* (Virtual Memory Address). Both addresses are mapped into a system address space.

Probably some hands-on examples would illustrate this better. Let's revisit the lost.c file you saved in Lab 2-1.

■ **Labs**   You will reuse the source file lost.c that you had from Lab 2-1. But this time you will link it to a different link script, as shown in Listing 2-5. Let's call this link script link_script02.x.

Listing 2-5. Link Script : link_script02.x

```
MEMORY
{
    ROM  : org = 0x200, len = 0x1000
    SRAM : org = 0x1300, len = 0x300
    DDR  : org = 0x1900, len = 0x200
}

SECTIONS
{

  my_code :
    { *(.text); } >ROM

  my_data :
    {*(.data);*(.rodata);} >SRAM AT>ROM

  .bss :
    {*(.bss);} >DDR

  other :
    {*(.*);} >SRAM
}
```

LAB 2-2: THE VMA/LMA/ROM IMAGE

The main difference between `link_script01.x` and `link_script02.x` is in the `my_code` and `my_data` sections. This time you want to store them in the ROM, but the data section has to be loaded into RAM after power-up, since we need to modify its content. You'll designate this special requirement by putting a `AT>ROM` at the end of the `my_data` section.

Compile `lost.c` again with this new link script:

```
$> gcc -fno-exceptions -c lost.c
$>
$> objcopy --remove-section=.comment lost.o
$>
$> ld -o my_output02.elf -T link_script02.x lost.o
$>
$> objdump -h my_output02.elf
```

Table 2-2 shows the output of `objdump`.

Table 2-2. *Objdump Output with Link Script 02*

Idx	Name	Size	VMA	LMA	File off	Algn
0	my_code	0000001b	00000200	00000200	00000200	2**2
		CONTENTS, ALLOC, LOAD, READONLY, CODE				
1	my_data	0000000c	00001300	0000021b	00000300	2**2
		CONTENTS, ALLOC, LOAD, DATA				
2	.bss	00000080	00001900	00001900	00000900	2**5
		ALLOC				
3	other	00000000	0000130c	0000130c	0000030c	2**2
		CONTENTS, ALLOC, LOAD, CODE				

As you can see, the `my_code` section has the same LMA and VMA value—both are equal to the starting address of ROM, which means the code will be stored in this ROM address and executed in place (XIP) from the same address. This makes sense since you expect the code to be read-only. (One exception to this rule is the self-modifying code, which will be visited in later chapters.)

However, also notice that the VMA and LMA of the `my_data` section now have different values. After power-up, the runtime library will copy the .data section from LMA (ROM address) to VMA (RAM address) so that it can be modified if necessary.

Since you are going to store `my_code` and `my_data` in ROM, it is imperative to see what the actual ROM image looks like. You can generate the ROM image using the following command:

```
$> objcopy -O binary my_output02.elf my_output02.bin
```

Open this raw binary image with Ultraedit (or any other tools that can view binary files):

```
Offset                      Content

00000000h:   55 89 E5 83 EC 10 C7 45 FC DD CC BB AA A1 08 13
00000010h:   00 00 C6 80 00 13 00 00 00 C9 C3 04 08 0F 10 17
00000020h:   2A 00 00 AA 9C 8B 7A
```

If you compare notes against the original source file (lost.c in Listing 2-3) and the link script in Listing 2-5, the following will come to your attention:

1. In lost.c, there is a global array called lost_candidates[]. The initial value of this array can be found at offset 0x1B. This marks the beginning of the .data section. And the offset 0x1B matches what you have in the link script. (0x1B = LMA of my_data ROM starting address = 0x21b - 0x200.)

2. There are two bytes of zero padding at the end of lost_candidates[] to make it align to the 32-bit boundary. This is done automatically by the compiler.

3. There is a constant integer in lost.c (John_Locke = 0x7A8B9CAA), which can be found in the raw binary image right after the lost_candidates[] array. This marks the .rodata section (read-only data).

4. There is also an uninitialized array of 128 bytes called jungle[128] in lost.c, which sits between array lost_candidates[] and constant John_Locke in the original source code. However, this uninitialized array is nowhere to be found in the raw binary image (The total size of my_output02.bin is less than 128 bytes). In fact, this uninitialized array will be in the .bss section. After power up, the runtime library will initialize the .bss section to zero in the RAM (VMA = LMA = 0x1900 in this case). The ROM image does not contain a .bss section.

5. Note that the raw binary image you've just generated is not functionally complete yet. The biggest missing piece is the runtime library (or any startup code). You need the runtime library to load the .data section into RAM and zero out the .bss section.

As you learned from Lab 2-2, VMA and LMA could be different for the ROM image. The .data section needs to be copied from ROM into RAM, and the .bss section needs to be initialized to zero at its corresponding RAM address. These are all done by the runtime library. The whole XIP flow for Lab 2-2 is illustrated in Figure 2-3.

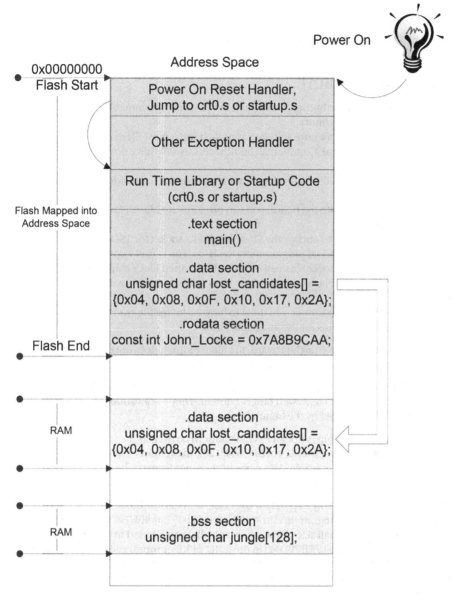

Figure 2-3. Execute in place (XIP) from a ROM image

Flash Memory

Most likely, your bootloader will be saved in Flash memory, which has its own idiosyncrasies.

Burn the Flash (PROM Program)

After the ROM image is produced, it has to be burned into an actual PROM (Programmable ROM) device. Most likely, the PROM device will be Flash memory.

You can program the Flash memory with a pure software approach since most Flash memory chips support the command set for read/write/erase. Commercial bootloaders usually have drivers for various Flash memory chips.

However, this now begs the question: At the beginning, how do you put the bootloader image itself into a blank Flash?

If your flash memory chip is soldered onto the board, you can program the blank Flash through the JTAG/SPI/I²C interface (the ISP approach).

Or you can solder a socket on the board to anchor the Flash (such as a socket for TSOP or QFP package), and program your blank flash through a universal programmer (Ref [10]). The socket approach has its merits in the early development stage. By putting a socket instead of a pre-determined Flash chip on the board, the developer can easily try out pin-compatible Flash memory chips from various vendors with different sizes without resorting to PCB rework.

Whether you use ISP or a universal programmer, you need to have an image file ready (like those files generated by objcopy.). There are a few options for the image file format:

- *Binary format*: Although it comes straightforward, the binary format does not contain checksums for itself, and most text editors don't handle binary files very well.

- *Motorola S Record, Intel HEX and Tektronix Extended HEX*: These are text file formats defined by different companies to represent binary content. They are all aimed to overcome the drawbacks inherited by the binary format.

- *Proprietary format*: If the flash chip is from a FPGA vendor to be used as a configuration device for FPGA (such as Altera EPCS64/128), the vendors sometimes define their own file format. For example, Altera has its own .pof format when you use Quartus software (Ref [30]) to program its Flash through JTAG.

Note that although JTAG is popular as a way to program the Flash in system, it is not the only interface. Instead many serial flash/EEPROM can be programmed in system through I²C or SPI. For embedded systems, a serial flash/EEPROM with very small size (a few bytes typically) can be used to store vendor ID and device ID (Ref [12]), while large serial flash/EEPROM (in the order of KB or more) can be used to store the whole boot image. Some microcontrollers, such as the 8051 in Ref [27], support serial Flash/EEPROM as a boot option. During power up, the boot image will be copied from serial Flash/EEPROM to on-chip RAM automatically by the microcontroller.

NOR Flash and NAND Flash

Although Flash memory was frequently mentioned in previous chapters, I did not provide details on the corresponding physical devices. I was reluctant to do so because there are many devils hiding in those details. And now is the time to let them out.

In terms of semiconductor technologies, there are two kinds of Flash memory being manufactured today: NOR and NAND. The semiconductor processes for them are beyond the scope of this book. In-depth information for Flash memory manufacturing can be found in Ref [15][16][17]. As far as developers are concerned, the difference between NOR and NAND are shown in Table 2-3.

Table 2-3. *Differences Between NOR Flash and NAND Flash*

	NOR Flash	NAND Flash
Random Read Access	Support	None/partial support
Density	Low	High
Cost per Bit	High	Low
Bad Block out of Factory	No	A small percentage
Read Interface	SRAM like	I/O, disk like
Write/Erase Time	Slow	Fast
Burst Access	Slow	Fast

If you take a close look at the datasheets of NOR Flash (Ref [15]) and NAND Flash (Ref [16]), you will notice that NOR Flash has separate address bus and data bus. In other words, the interface of NOR Flash is pretty close to that of SRAM's. In fact, NOR Flash is good at random reading, and it does not carry any bad block out of factory. This makes NOR Flash suitable for storing bootloaders. On the other hand, NAND Flash's interface is not so straightforward. The developers have to jump through hoops to get the job done. So the conventional wisdom is to use NOR Flash for code storage (such as bootloader or OS image) while have NAND Flash as mass data storage.

However, the ecosystem of Flash world has gradually changed over the past ten years. With the advance in MLC (Multi-level Cell) technology and the fine tuning of SLC (Single Level Cell), NAND Flash has greatly reduced its cost per bit through significantly higher density. In the meantime, CPU manufacturers and NAND Flash manufacturers have worked jointly to support direct booting from NAND Flash, mainly through code shadowing, as illustrated in Figure 2-4.

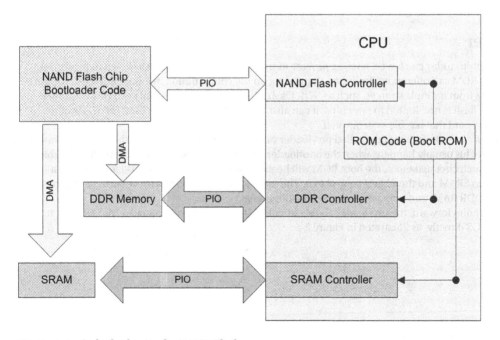

Figure 2-4. *Code shadowing for NAND Flash*

1. To do code shadowing, the CPU's memory controller has to support a NAND interface. Take TI OMAP 35x (Ref [28]) for example. The OMAP 35x family has a GPMC (General Purpose Memory Controller) interface. It is a bus protocol for TI OMAP processor to talk to external memories. A similar bus protocol called EIM (External Interface Module) can be found on the NXP iMX processor (Ref [29])) that could talk to NOR Flash, SRAM, or NAND Flash, depending on the configuration.

2. The CPU has to have a minimal amount of ROM code that can copy the bootloader from NAND Flash to RAM. And the physical NAND Flash chip that interfaces with the CPU should be supported by such ROM code (Boot ROM). Bootloader can be executed in place (XIP) after it is copied into RAM. This code copying process is usually called *code shadowing*.

3. Since NAND Flash is good at burst reading, DMA could be used to reduce the copying time.

4. Since NAND Flash inherently has bad blocks out of factory, ECC code needs to be in place to guarantee data integrity. Some CPUs, such as OMAP 35x, have a built-in ECC calculator to help the number crunching. Meanwhile NAND Flash manufacturers have also done some work on their end. For examples, manufacturers now supply some models (Ref [17]) of NAND Flash with the first block always valid so that no ECC is required for the first block. Bootloader thus can be stored in this block.

While NAND Flash is bettering itself, new processes are also introduced to NOR Flash (Ref [18][19]) to boost its single chip capacity over 512Mbits. Meanwhile, different flavors of memories now can be put into the same chip thanks to the MCP (Multi-Chip Packaging) and POP (Package on Package) technologies (Ref [20]). So my suggestion is to shop around for new designs to get the best Flash solution in terms of cost and complexity.

Pre-Loader

In reality, SoC chips today can have a booting process more complicated than what's shown in Figure 2-4, and using boot ROM can offer more flexibility. Depending on the configuration, the boot ROM can support booting options from multiple sources, such as NOR Flash (XIP), NAND Flash, serial port, USB, Ethernet, etc. The NAND flash is not limited to raw chips; it can also be a MMC (Multi Media Card) or SD (Secure Digital) memory card that has file systems on it.

Sometimes an intermediate stage called pre-loader might be introduced between the boot ROM and the bootloader. This usually happens when the bootloader has a footprint much bigger than the available SRAM. Under such circumstances, the boot ROM will begin loading a lightweight bootloader—called a *pre-loader*—into SRAM and then hand control to it. The main job of pre-loader is to initialize hardware, especially the DDR RAM, and prepare for the next loading stage.

From that point forward, the pre-loader could either load the full-blown bootloader into DDR, or it could load the OS directly, as illustrated in Figure 2-5.

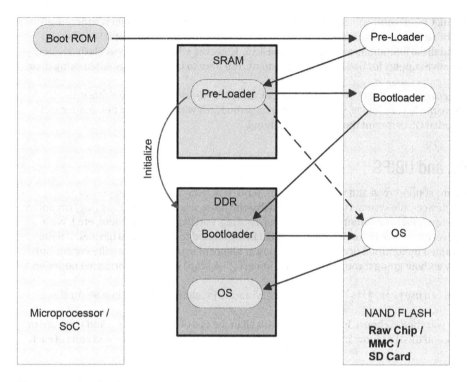

Figure 2-5. *Pre-loader*

Some bootloaders, such as U-Boot, offer options to make a pre-loader in their building and customization process. In U-Boot, this pre-loader is called SPL (Secondary Program Loader, Ref[24]), and it is in fact a lightweight bootloader with a small footprint.

Flash File System

Flash devices are different from hard disks in that Flash devices need to go through an extra erasure cycle prior to a write cycle. The block sizes of Flash devices are much bigger than their hard disk counterparts. Typical block size for Flash devices can be somewhere between 16KB to 512KB, while the sector size of a hard disk is usually 512 bytes. Also each block on a Flash device can only survive a limited number of P/E (Program/Erasure) cycles, which means load balancing (wear-leveling) is necessary for Flash devices. For NAND flashes, bad block management is also a must-have because NAND flash would carry a certain amount of bad blocks out of factory.

Due to such quirkiness, Flash devices' file systems would be different from those used on hard disks. Commonly used flash file systems are discussed in the following sections.

RomFS, CramFS, and SquashFS

These are all read-only file systems. Folders that contain your object modules and executables are good candidates for these file systems, and they take less time to mount during system boot up.

Among these file systems, the RomFS code base is the smallest, and so is its memory consumption. However, due to its simplicity, it does not support compression.

On the other hand, CramFS and SquashFS are both compressed file systems. CramFS is widely used by embedded Linux, but it carries some limitations on file size, maximum number of user IDs and group IDs, etc. (Ref [21]). And these limitations are lifted by SquashFS. However, when it comes to Flash devices, neither of them has native support for bad block management. You have to take extra care when using them on NAND flashes.

Under Linux, you can use genromfs[2], mkcramfs,[3] and mksquashfs[4] to create images for those file systems. After you have made your image, you can burn it onto the Flash with the help of bootloader and configure your embedded OS to mount these image partitions.

JFFS2, YAFFS2, and UBIFS

These Flash file systems support read and write operations. Since it is always a lengthy process to program/erase Flash devices, file systems that support Flash write must also have journaling (log) capability in addition to all the requirements mentioned earlier (wear-leveling, bad block management, etc.). With journal file systems, every write operation is logged before it is committed to the physical devices. If write operations are interrupted by common occurrences like power failure or system crash, the file system can be remounted quickly without going through a recovery process (disk scan) or leaving corrupted nodes on the flash.

Under Linux, you can use mkfs.jffs2[5], mkyaffs2image,[6] and mkfs.ubifs[7] to create images for those file systems.

As for which one you should choose for your design, it's all in the eye of the beholders, and there are no carved-in-granite rules on this. For more information, refer to Ref [21][22][23] for the pros and cons of each file system.

Summary

To summarize, bootloader has the following important traits:

- It is the first program to be executed after power on.

- It has to find a way to load itself into RAM or to be executed in place (XIP).

- It loads image files into memory and executes them. Image files are composed of multiple sections, like code, data, bss, etc.

- It is saved in Flash memory. There are two kinds of Flash: NOR and NAND.

And for any code (including bootloader) to be executed, you need CPU, memory, and bus. The next chapter discusses them in detail.

---

[2]genromfs package needs to be installed if it is not readily available.
[3]It could also be mkfs.cramfs on some Linux systems.
[4]squash-tools needs to be installed.
[5]The mtd-utils package needs to be installed to make the make.jffs2 command available.
[6]This command needs to be built from source code if it is not readily available.
[7]Kernel source code and mtd-utils source code are needed to build the ubifs utilities.

References

1. ARM926EJ-S (Rev r0p5) Technical Reference Manual, ARM Limited, June, 2008
2. TMS320VC5402 Fixed-Point Digital Signal Processor (SPRS079G), Texas Instruments Incorporated, October, 1998
3. TMS320VC5402 and TMS320UC5402 Bootloader (SPRA618B), Texas Instruments Incorporated, November, 2004
4. RedBoot (http://sourceware.org/redboot/)
5. U-Boot (http://www.denx.de/wiki/U-Boot/WebHome)
6. *Code Composer Studio User's Guide* (SPRU328B). Texas Instruments Incorporated, November, 2000
7. Tool Interface Standard (TIS) Executable and Linking Format (ELF) Specification, Version 1.2. TIS Committee, May, 1995
8. Monk TV show, USA Network (http://www.usanetwork.com/series/monk)
9. µClibc, http://www.uclibc.org
10. Xeltek SuperPro Programmer, Xeltek Inc. (http://www.xeltek.com)
11. Phase Locked Loop (ALTPLL) Megafunction User Guide, Altera Corporation, November, 2009
12. 24AA00/24LC00/24C00, 128 bit I²C EEPROM, Microchip Technology, Inc., 2007
13. *Digital Integrated Circuits—A Design Perspective (2nd Edition)*. Jan M. Rabaey, Anantha Chandrakasan, and Borivoje Nikolic. Pearson Education, Inc., 2003
14. *Silicon VLSI Technology—Fundamentals, Practice and Modeling*. James D. Plummer, Michael D. Deal, Peter B. Griffin, Prentice Hall, Inc., 2000
15. 3 Volt Intel StrataFlash Memory—28F128J3A, 28F640J3A, 28F320J3A (x8/x16), Intel Corporation, 2001
16. NAND Flash Memory (MT29F2G08AABWP/MT29F2G16AABWP, MT29F4G08BABWP/ MT29F4G16BABWP, MT29F8G08FABWP), Micron Technology, Inc., 2004
17. NAND16GW3F4A, 16-Gbit (2 x 8 Gbits), two Chip Enable, 4224-byte page, 3V supply, multiplane architecture, SLC NAND Flash memories, Numonyx, B.V., November, 2009
18. MirrorBit Technology: The Foundation for Value-Added Flash Memory Solutions. Spansion MirrorBit Technology Brochure. Spansion LLC, 2008
19. S29WS-P, MirrorBit Flash Family, S29WS512P, S29WS256P, S29WS128P 512/256/128 Mb (32/16/8 M x 16 bit) 1.8 V Burst Simultaneous Read/Write MirrorBit Flash Memory, Spansion Inc., 2008
20. S72WS-N Based MCP/PoP Products, 1.8 Volt-only x16 Flash Memory and SDRAM on Split Bus 256/512 Mb Simultaneous Read/Write, Burst Mode Flash Memory 512 Mb NAND Flash 1024 Mb NAND Interface ORNAND Flash Memory on Bus 1 512/256/128 Mb (8M/4M/2M x 16-bit x 4 Banks) Mobile SDRAM on Bus 2, Spansion Inc., 2007
21. *Building Embedded Linux Systems, 2nd Edition, Concepts, Techniques, Tricks, and Traps*. Karim Yaghmour, Jon Masters, Gilad Ben-Yossef, and Philippe Gerum. O'Reilly Media, August 2008
22. *Anatomy of Linux Flash File Systems, Options, and Architectures*. M. Tim Jones, Emulex Corp., May, 2008
23. UBIFS—UBI File-System (http://www.linux-mtd.infradead.org/doc/ubifs.html)
24. Altera SoC Embedded Design Suite User Guide, Altera Corporation, December, 2014
25. "Why I don't like printf()." Erich Styger, MCU on Eclipse, http://mcuoneclipse.com/2013/04/19/ why-i-dont-like-printf/
26. Secrets of "printf," Professor Don Colton, Brigham Young University Hawaii
27. EZ-USB Technical Reference Manual, Version 1.2, Cypress Semiconductor Corporation, 2005
28. OMAP 35x Application Processor, Technical Reference Manual, Literature Number: SPRUF98L, Texas Instruments Incorporated, November, 2010
29. i.MX 6Dual/6Quad Applications Processor Reference Manual, Document Number: IMX6DQRM, Rev. 3, Freescale Semiconductor, Inc. July, 2015
30. *Quartus Prime Standard Edition Handbook,* Altera Corporation. May, 2015
31. QDR™-II, QDR-II+, DDR-II, and DDR-II+ Design Guide, AN4065, Cypress Semiconductor Corporation, November, 2007
32. DDR3 SDRAM (MT41J256M4—32 Meg x 4 x 8 banks, MT41J128M8—16 Meg x 8 x 8 banks, MT41J64M16 - 8 Meg x 16 x 8 banks) Datasheet Rev I, Micron Technology, Inc., February, 2010
33. CIO RLDRAM II (MT49H32M9—32 Meg x 9 x 8 Banks, MT49H16M18—16 Meg x 18 x 8 Banks, MT49H8M36 – 8 Meg x 36 x 8 Banks) Datasheet Rev N, Micron Technology, Inc., May, 2008

CHAPTER 3

Inside the CPU

Help Desk: Double-click on "My Computer"
User: I can't see your computer.
Help Desk: No, double-click on "My Computer" on your computer.
User: Huh?
Help Desk: There is an icon on your computer labeled "My Computer". Double-click on it.
User: What's your computer doing on mine?

—*How to be a Good Teacher,* by Rupal Jain

As you've seen, the CPU is the command and control center of each and every embedded system. Ever since the time when the first computer was conceived, many great minds have dedicated the better part of their lives to new CPU architectures. The topic of CPU architecture is too pandemic to be discussed here. Inquisitive readers would be better served by Ref [1][2] and other good books on this topic. Fortunately, as far as embedded systems are concerned, only certain parts of it matters to your development practice. This chapter tries to glean these pieces.

Von Neumann Architecture and Memory Barrier

In terms of CPU architecture, there are two main camps: Von Neumann Architecture and Harvard Architecture. The family tree of the Von Neumann Architecture goes back to the days when the first digital computer was born, and the essence of the Von Neumann Architecture is "stored-program". Basically, this means data and code share the same address space, and they are stored in the memory without explicit effort to distinguish them, as shown in Figure 3-1.

© Changyi Gu 2016
C. Gu, *Building Embedded Systems*, DOI 10.1007/978-1-4842-1919-5_3

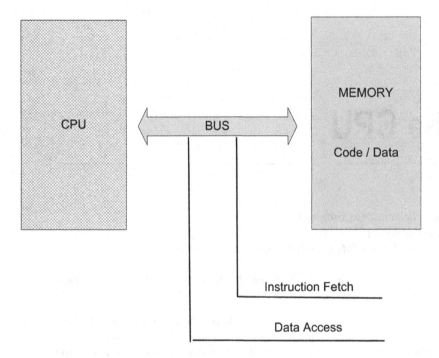

Figure 3-1. *Von Neumann Architecture*

Thanks to such indiscrimination of code and data, the following engineering practice became possible:

- Self-modifying code. Under the Von Neumann Architecture, It is possible to generate instructions dynamically as data, and then execute them as code. However, such practice is largely discouraged in the embedded system development because it would complicate the matter without obvious benefit. Unless you believe you have a strong case for your particular application, stay away from self-modifying code.

- Since there is no explicit difference between data and code, bootloader can initially treat user images as pure data and load them into the memory. Afterward, these images can be executed as instructions. A typical case would be your PC. Every time you turn on your PC, the BIOS on the motherboard will start to load the Windows image (Excuse me for this presumption.) from hard disk into your RAM and Windows will take over your computer.

So if you decide to do something similar to the behavior of bootloader, the pseudo-code may look like Listing 3-1.

Listing 3-1. Attempt to Turn Data into Code

```
...

(1)    initialization

(2)    load code into memory address #A

(3)    Jump to #A to execute the new code
```

However, things are actually more complicated than this:

1. Modern CPUs all have pipeline structure. Instructions could be pre-fetched or speculatively executed. To be on the safe side, before Step 3 is executed in Listing 3-1, it is better to flush the pipeline.

2. Many modern-day CPUs are in fact hybrids of the Von Neumann Architecture and the Harvard Architecture. A typical case is to use a separate I-Cache (Instruction Cache) and D-Cache (Data Cache) while keeping a unified address space for memory. After Step 20 in Listing 3-1, D-Cache might contain something that is actually code. They need to be invalidated.

3. The compiler might do some optimization that you are unaware of.

For the scenarios 1 and 2 mentioned above, extra steps have to be taken to guarantee the CPU's normal behavior. In fact, *whenever code is treated as data, a sequence of special instructions must be used to ensure consistency between the data and the instruction streams processed by processor* (Ref [10]). Such special instructions are usually called the "memory barrier". The memory barrier instructions are unique to each type of CPU. For uni-processor systems, they are often composed of instructions to flush the pipeline or invalidate the cache.

Note that the memory barriers are *hardware* memory barriers. They should be applied to make the CPU hardware behave normally. However, compilers would also do optimizations to change the order of load and store. When the data is going to be executed as instructions, such compiler optimization could cause unintended consequences. Fortunately, most mainstream compilers offer memory barrier macros/functions (such as _ReadWriteBarrier in Microsoft VC++) to prevent moving memory instructions across the memory barrier. So to be on the safe side, it is suggested for developers to insert a hardware memory barrier and a compiler memory barrier between Steps 2 and 3 in Listing 3-1, which will make the process look like Listing 3-2.

Listing 3-2. Apply Memory Barrier

```
...

(1)    initialization

(2)    load code into memory address #A

(3)    >>> Compiler Memory Barrier Macros/Functions
(4)    >>> hardware memory barrier

(5)    Jump to #A to execute the new code
```

Note that this discussion is based on the assumption of a uni-processor system. For dual-processor (or multi-processor) systems that have the shared memory, things are even more complicated, as modern-day CPUs may often execute instructions out of order to boost performance. And memory barriers have to be carefully applied to all processors to guarantee the proper data exchange through shared memory.

Harvard Architecture, Modified Harvard Architecture

Different from the Von Neumann Architecture, the Harvard Architecture uses two separate physical spaces (two address spaces) to store code and data, as shown in Figure 3-2. The plus side of such an arrangement is that instruction fetching and data access can happen at the same time, which greatly increases bus throughput. Thus for the same clock frequency, a CPU of the Harvard Architecture usually has better performance (speed) over its Von Neumann counterpart, and is thus widely adopted by many digital signal processors, like TI's C55x (Ref [4]) and ARM's ARM9 (Ref [1]). However, like all digital designs, the Harvard Architecture obtains the speed advantage at the expense of a larger area[1].

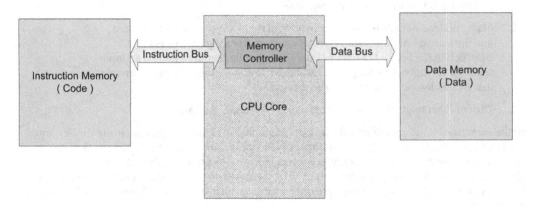

Figure 3-2. *Harvard Architecture*

SIDE NOTES

For CMOS circuits[2], the dynamic power is estimated by

$$P_{dynamic} = \alpha \cdot C \cdot V_{DD}^2 \cdot f$$
Equation 3-1 (Ref [5])

where α is the activity factor, C is the load capacitance, V_{DD} is the supply voltage, and f is the clock frequency. In this sense, the relatively low clock frequency of the Harvard Architecture is good for dynamic power[3]. However, as CMOS manufacturing processes advance into deep micron territory, high leakage has made static dissipation (quiescent power) on the same order of dynamic dissipation, which works in disfavor of the Harvard Architecture's larger area. Thus you should be careful not to jump to any immediate conclusions when you judge power dissipation based on CPU architectures.

For general-purpose microprocessors, Intel's 8051 (its CMOS counterpart is 80C51) epitomizes what a typical Harvard Architecture ought to be. It has two separate address spaces for code and data. Code is stored in ROM and executed in place (Ref [3]). Although it is only an 8-bit processor, it is still extensively supported by many IC vendors today.

However, for the Harvard Architecture, such "XIP from ROM" solution is not suitable for everybody. The separation of code space and data space actually puts a strain on the bootloader. Some CPUs solve this by having a special load option that loads code from Flash/EPROM into code space during boot time. Others tackle this by providing instructions to move things between code space and data space if necessary.

---

[1]The trade-off between area and time will be covered in later chapters for FPGA design.
[2]Most digital chips manufactured today (including most CPUs) are in the CMOS process.
[3]Power dissipation is a very involved subject that is beyond the scope of this book. Readers can refer to Ref [5][9][10] for more information.

The latter one is sometimes called the Modified Harvard Architecture. Since the Modified Harvard Architecture raises the possibility of treating code as data, a memory barrier can be used to make the CPU behave properly.

To explicitly distinguish between code and data, some CPUs with a Harvard Architecture will use a flag called NX bit (No-eXecute bit) to mark memory regions dedicated to data. However, this idea was soon borrowed by the Von Neumann camp for security reasons, which is often used to prevent virus code disguised as data from being executed.

Also note that for digital signal processors of the Harvard Architecture, they usually carry internal memories (separately for code and data) that are close to the CPU core. For example, ARM9 has something called TCM (Tightly Coupled Memory) for code and data. These TCMs are independent of I-Cache/D-Cache, and can be accessed with low latency and high throughput. In addition, external memories, like DDR, can also be accessed by ARM9 through an AHB bus. Developers should always squeeze time critical code (hard real-time) into internal memory while keeping the non-real-time part in external memory through the proper setup in the link script.

CISC and RISC

The holy war between CISC and RISC started long before I graduated from engineering school. Arguments from both sides galore are on the Internet, and conclusions can be found in Ref [2]. I just want to offer my two cents from the perspective of a FPGA designer/firmware engineer.

As I will reiterate in later chapters for FPGA, there are only three major ways to improve the throughput of a digital design: pipeline, parallel processing, and ping-pong buffer. In the early days when CPUs were being designed, memory size was small and memory prices were high. Clock frequency was relatively low though. So naturally when people were designing instruction sets, they tended to make each instruction do as much as it could, which led to instruction set with the following characteristics:

- Instructions become irregular. Due to the multitude of addressing modes, some instructions were longer than others. Complicated instructions took more clock cycles to complete than simple instructions.

- The mushroom of complicated instructions amounts to a perplexing datapath[4] when they are implemented in hardware.

Computers of such instruction sets are called CISC (Complex Instruction Set Computer). And I believe CISC designers later found themselves having a difficult time pushing for pipeline and parallel processing, thanks to the cumbersomeness inherited instruction set. The inefficiency of pipeline implementation and the large granularity at the instruction level made it very hard to boost clock rate or throughput.

Thus CPU designers started to move away from the CISC-style CPU in favor of more concise instruction sets. RISC (Reduced Instruction Set Computer) designs become more and more popular. RISC design advocates a regularity in instruction length and format. Except for memory load/store, most instructions only operate on registers and immediate numbers. Regularity makes pipeline more efficient, and the average lightweight instruction means smaller granularity, which is good for exploiting parallelism at the instruction level.

At this point, a safe bet is that 90% of the CPUs on the market are RISC implementation. Here I've used the phrase "RISC implementation" because some CPU families, like x86 and 8051, include a strong CISC legacy. Due to the astronomical prevalence of legacy software, it is incumbent on them to be binary-compatible with existing code bases. So they choose to support CISC instructions at the software level and translate CISC instructions into internal RISC operations through hardware (Ref [1][6]).

---

[4]Datapath and controller will be covered in later chapters for FPGA.

The popularity of RISC CPUs made pipeline a universal reality. Pipelines can introduce certain awkwardness on the software side. Developers might have to do some optimization in the following two aspects:

1. *Delay Slot*

   Pipeline favors a constant flow of instructions. And we all know this is not always the case given the complicated nature of software. Load instructions and branch instructions are the two main culprits that can stall the pipeline. Due to the relatively slow nature of external bus/memory as opposed to CPU main clock, the data that is being loaded may not be readily available at the next pipeline stage. So the instructions immediately following the load cannot operate on this data right away, as shown in Listing 3-3 (such dark period after load/branch is called a *delay slot*).

Listing 3-3. Pipeline Data Hazard

```
regB <= regB + 1

load register : *(mem addr) => reg A

regA <= regA + 1 (cannot be done by pipeline!)
```

An easy way to fix this is to put NOP instructions after the load, as shown in Listing 3-4.

Listing 3-4. Fill NOP in the Delay Slot

```
regB <= regB + 1

load register : *(mem addr) => reg A

NOP (assume there is one delay slot for load)

regA <= regA + 1
```

To fine-tune the performance and better utilize the delay slot, developers can put instructions that do not depend on the load/branch into this slot, as shown in Listing 3-5.

Listing 3-5. Fill in the Delay Slot with Unrelated Instruction

```
load register : *(mem addr) => reg A

regB <= regB + 1 (assume one delay slot for load)

regA <= regA + 1
```

Although this example is for load instructions, the same can be said for branch instructions. Compilers can support delay slot optimization automatically or through compiler switch. If assembly is used for programming, as many DSP programmers would prefer, such tricks have to be applied manually.

2. *Loop Unrolling*

   Loop unrolling is not entirely confined to RISC CPUs. However, loop unrolling will open more doors for optimization when pipeline and delay slot are involved. As shown in Listing 3-6, assuming one clock cycle of loop overhead and one delay slot for load instruction, it takes roughly 64 cycles to finish the main loop (16 cycles of loop overhead, three cycles of execution time each loop).

Listing 3-6. Loop without Loop Unrolling

```
regA <= start address

regC <= 0

for (i = 0; i < 16; ++i) {

    load : *(regA++) => regB;
    NOP;
    regC <= regC + regB;

}
```

Now if you tried loop unrolling by two fold, you would have something like Listing 3-7.

Listing 3-7. Loop Unrolling

```
regA <= start address
regAA <= regA + 8;

regC <= 0;
regCC <= 0

for (i = 0; i < 8; ++i) {

    load : *(regA++) => regB;
    load : *(regAA++) => regBB;
    regC + regB => regC;
    regCC + regBB => regCC;

}
regCC + regC => regC;
```

After loop unrolling, the main loop takes 40 cycles to finish (eight cycles of loop overhead and four cycles of execution time for each loop). The saving comes from reduced loop overhead as well as better optimization for the delay slot.

Lastly, to avoid any possible confusion, be advised that the term CISC/RISC is mainly about instruction sets and corresponding hardware implementation, while the concept of Von Neumann/Harvard in previous sections deals largely with memory bus layout. As shown in Table 3-1, any combination of CISC/RISC and Von Neumann/Harvard is possible.

Table 3-1. *CISC/RISC and Von Neumann/Harvard*

	Von Neumann	Harvard
CISC	Intel 80386	Intel 8051
RISC	Intel Pentium II[5]	ARM926

---

[5]Strictly speaking, it is a hybrid of Von Neumann and Harvard, with a unified memory address space and separate L1 caches for instruction and data.

SIMD/VLIW

As I mentioned, there are only three major ways to improve the throughput of digital design: pipeline, parallel processing, and ping-pong buffer. Thus it's natural to want to add multiple PEs (processing elements) to the hardware. Ideally, if you place *n* PEs in parallel and keep them all busy in each instruction cycle, you can increase the throughput by a factor of *n*.

CPUs using this approach are called SIMD or VLIW processors. (My understanding is that processing elements in SIMD all perform the same function while VLIW does not have this restriction. Consult academic pedants for the details.) With this approach, the CPU designer is actually shifting some of the optimization burden to the software side. To take advantage of such parallel architecture, the programmer has to do the optimization manually or rely heavily on the assistance of compiler. As an interesting side note, there is an optimization trick called *software pipeline* that's unique to the VLIW architecture. It is kind of like loop unrolling with a VLIW flavor. I leave it to the readers to explore further.

CPU Debugging

For engineers, actions always speak louder than words. Our ultimate mission is to keep the whole system spinning the way we anticipated. However, like any systems built by human beings, imperfection is part of the daily life. So when things go awry, you need handy tools beside poking and prodding to figure out what's going on inside the CPU.

CPU Simulator

One approach is to use software to mimic the behavior of the target CPU, and such software is called a *CPU simulator*. The plus sides to using a simulator is that:

- You don't have to have the target CPU physically available. All you need is a host machine (such as a PC) to run the simulator. So hardware costs can be kept low during development.

- Software engineers can have an early start when the hardware is still being developed. As discussed in later chapters, the concurrent workflow of the hardware (including FPGA/AISC) and software teams is the norm for embedded system development. Simulators can help software engineers catch the majority of bugs without involving hardware.

- Breakpoints can be set up freely. Register content can be fathomed without restriction.

The drawbacks of simulators are also conspicuous:

- They run slow. By slow, I mean a factor of 1,000 or more. So the simulation time could become a big headache for large projects.

- Asynchronous events cannot be faithfully reflected in a simulator. Because of the slow code execution, many real-time events, like interrupt from the peripherals, may not be consistent with the actual hardware behavior.

So it would be better if you could have something that helps you on the real hardware.

ROM Monitor

One way to alleviate the pain of bug-bite is to beef up the bootloader—give it better debugging capability, like reading/writing the memory and registers. As mentioned, such a bootloader is sometimes called *ROM Monitor*. A more powerful ROM Monitor would even support software breakpoint, making it look a lot like the debugger you have on PC/Windows. For many low-end microcontrollers, such as the 8051, ROM Monitor provides a low-cost solution for debugging (Ref [7]).

However, as you can imagine, ROM Monitor has its limitations:

- ROM Monitor is a piece of software, just like the software it is trying to debug. Inevitably, it will consume memory and other hardware resources. For example, most ROM Monitors need a debug console to interact with user, which usually requires one UART port dedicated for this purpose.

- Since ROM Monitor is also a piece of software, its memory space could be corrupted by other software, or it might conflict with other software on hardware resources.

- ROM Monitor has limited control over the software it is trying to debug. For example, if the buggy software is accidentally trapped in an ISR with an endless loop (assume UART interrupt is disabled in this case), the debug console would stop responding to user input, and you might have a hard time pinning down the PC counter value in this circumstance.

- By the way, who is going to help those folks who develop the ROM Monitor? Or the better question might be: how do you debug the ROM Monitor? Without any hardware assistance, this really becomes a chicken and egg dilemma. As the complexity of CPU grows exponentially, it becomes more and more challenging to write a ROM Monitor anew without any hardware assistance. And it is not a small undertaking to even port a generic bootloader, like RedBoot (Ref [2]) or U-Boot (Ref [3]), to a new hardware platform.

Come to think of it, ROM Monitor is indeed a software solution for software debugging. With the complexity of hardware and software balloons, engineers start to wish for a better hardware tool to probe inside the CPU. ICE is one of the answers to such calls.

Introduction to ICE (In Circuit Emulator)

Engineers' dark days usually begin from the workbench when the reality and the simulator are telling different stories. Fortunately, there is a piece of equipment called ICE (In Circuit Emulator) that can rescue engineers from these nightmares. There are two types of ICEs in this world: *non-JTAG ICE* and *JTAG-based ICE*. Analogically speaking, ICE can be viewed as a simulator that runs in hardware.

Non-JTAG ICE

In the good old days when CPUs were slow and less complicated, ICE was built in the following way (see also Figure 3-3):

1. The PCB board was built with a CPU socket on it. During development, the CPU socket was hooked to the ICE box through a cable. Inside the ICE there was a target CPU, and this CPU was specially built for debugging purposes, which was called bond-out CPU.

2. Bond-out CPU has many of its internal status wired out (or muxed-out) to the outside world for debugging purposes. If there is an internal ROM, such ROM will be replaced by external RAM instead. All the bus activity between CPU and external RAM will be captured by the ICE hardware.

3. Because there is no internal ROM for the bond-out CPU, all the instruction fetch and memory data access is reflected on the external bus and thus can be closely monitored by ICE.

4. Since the ICE has access to the internal registers through bond-out, it can modify the CPU internal status for debugging purposes as well.

5. When development is finished, a production CPU can be plugged into the CPU socket. Or if the socket and production CPU have the same footprint, production CPU can be soldered onto the PCB to replace the socket in the release version.

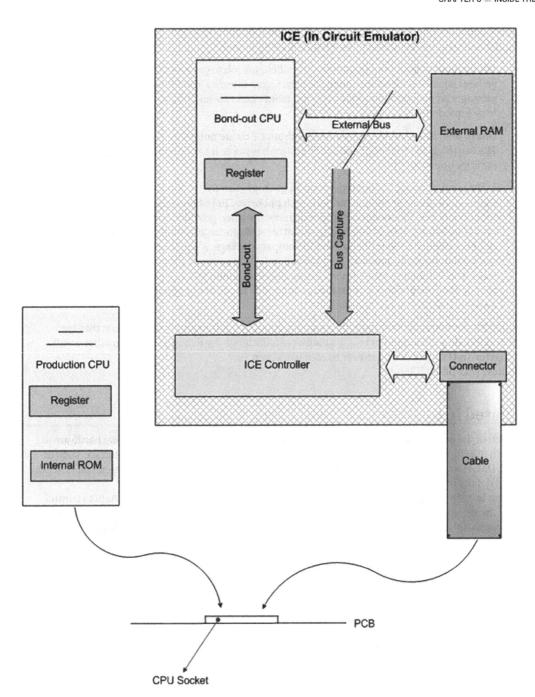

Figure 3-3. *Non-JTAG ICE, production CPU, and CPU socket*

However, with the explosive increase of bus speed and CPU clock frequency, the semi-conductor industry's glitz and glamour has pushed Non-JTAG ICE into a corner:

- The increase of bus speed has made it a challenging job to capture the bus activity without negatively affecting signal integrity. Signal integrity also becomes more pronounced for bond-out CPUs when the main clock rate and CPU complexity both have exploded.

- Due to their low-production volume, bond-out CPUs are notoriously expensive. The fast pace of CPU design and development has made it hard to procure the spare parts for bond-out CPUs.

- In the good old days, CPUs were offered in packages like DIP or PLCC. It is relatively easy to make a socket compatible with such packages. To reduce parasitic lead inductance, new CPUs with faster speed grades are often provided in more compact packages, like TQFP or BGA. Making socket for such packages usually means pin extension, which defeats the purpose of compact package. A workaround is to have two versions of PCBs, one with the production CPU in compact package, and the other with ICE socket. However, such a workaround would give rise to extra R&D cost. And the PCB with ICE socket may not be able to run at top speed due to added parasitic inductance and loading from ICE.

In hindsight, CPU designers are partially responsible for ICE's misfortune. Initially when they were designing the CPU, they did not put in enough hardware to facilitate the debugging task. In other words, DFT[6] (Design for Test) should be the answer to ease ICE's pain.

So comes JTAG-based ICE.

JTAG-Based ICE

As mentioned in the previous section, CPU designers are expected to design extra debugging hardware to interface with ICE. In fact, some CPU manufacturers have defined their own debugging interface, such as Motorola's BDM. However, over the years, microprocessor industry has pretty much converged on JTAG as the standard debugging interface.

JTAG was initially designed for IC I/O test in PCB assembly. It uses a boundary scan chain to control and capture the in/out of every IO Pad on an IC, as shown in Figure 3-4. BSCs (boundary scan cells) are inserted between the IO pad and the internal logic. Conceptually, BSC can be viewed as a MUX.

---

[6]DFT will be briefly covered in later chapters.

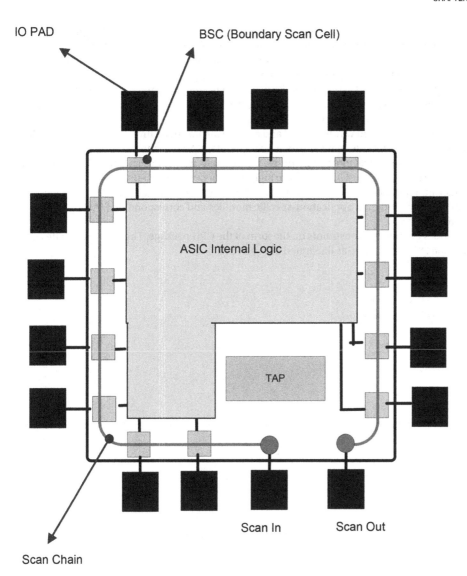

IO PAD

BSC (Boundary Scan Cell)

ASIC Internal Logic

TAP

Scan In

Scan Out

Scan Chain

Figure 3-4. *Boundary scan*

During normal operation, BSC is in pass-through mode, so there is no impact on the ASIC function. Under testing mode, 0/1 bit streams can be clocked into the BSC to set it high or low. Output from the ASIC can also be captured by the BSC and clocked out through the scan chain. If there are shorts to ground or power rail, they will be detected by clocking in various test vectors through the scan chain. Such boundary scan practice has been standardized in JTAG through the definition of connector pins, TAP (Test Access Port), etc. A description language called BSDL (Boundary Scan Description Language) was also defined to standardize the characterization of JTAG devices (see Figure 3-4).

As you can see, JTAG can be a good way to control and capture the in/out of any digital module. It didn't take long for IC designers to recognize its value as a debugging interface. So designers start to put JTAG around not only IO pads but also crucial internal modules. For example, in SoC design, the CPU core vendor may expose several JTAG chains for things like CPU core logic, memory systems, co-processors, etc., and leave some chain index number unassigned in the TAP controller for external modules. IC designers who adopt the CPU core could thus add their application-specific modules and corresponding JTAG chain using these unassigned index numbers.

JTAG-based ICE does not put any constraints on the form of the CPU package. The only real-estate cost of JTAG-based ICE is the JTAG connector, as illustrated in Figure 3-5.

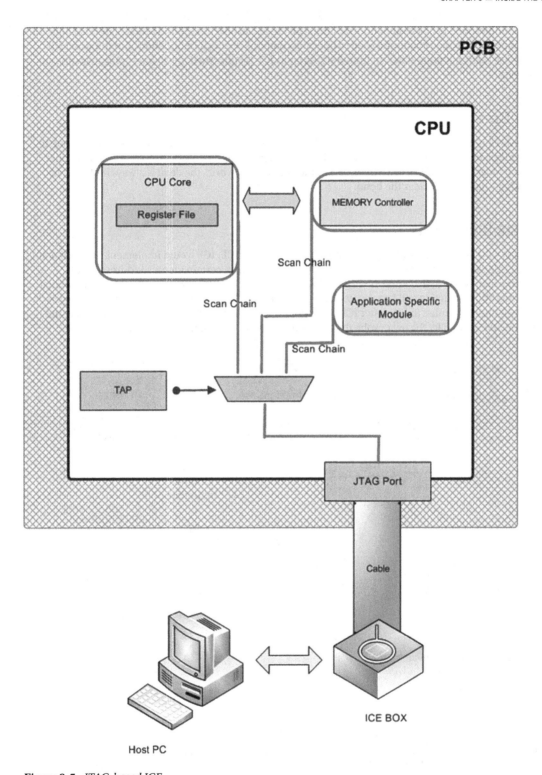

Figure 3-5. *JTAG-based ICE*

For completeness of discussion, note that the scan chain can also be inserted into IC's internal logic for DFT purposes. These scan chains are used to inject test patterns during IC fabrication, and the scan cells[7] used in these scan chains are different from BSC. As far as ICE is concerned, these internal scan chains are transparent.

Using Breakpoints

Anyone who has ever used an ICE will know first-hand that using *breakpoints* are an indispensable tool. In fact, any magic box calling itself an ICE would offer breakpoint capability. Engineers are so used to it that they take it for granted. But like everything else in the engineering world, the devil is always in the details. Here is what's happening under the hood.

Software Breakpoints

If the target CPU does not have any special hardware to assist the ICE, ICE would implement the breakpoint through a software approach. When the user sets up a breakpoint on certain memory addresses, ICE replaces the instruction at that address with an SWI instruction. When PC (Program Counter) reaches that address, this SWI instruction will be executed and the corresponding ISR prepared by ICE will be triggered[8]. Control will thus be handed to ICE. After ICE gets the control, it will restore the original instruction at the breakpoint address and let the user probe the CPU status. Figure 3-6 demonstrates such an approach.

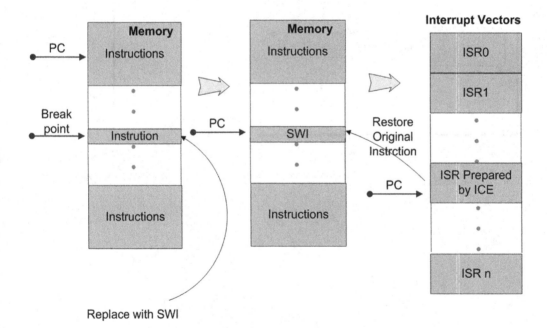

Figure 3-6. *Software breakpoints*

---

[7]There are three main types of internal scan chain style: multiplexed flip-flop, clock scan, and LSSD. Refer to Ref [8] for details.

[8]There are many ways to let software trigger interrupts. Using SWI instruction is one of them. Illegal instructions or errors like "divide by zero" can also trigger interrupts. Some books call those interrupts triggered by software *software traps*.

The plus sides of software breakpoints is that:

- There is no hardware cost associated with using software breakpoints, which means there is no limit to the number of software breakpoints that can be set simultaneously.

- Since software breakpoints do not need any hardware support, they are not confined to ICE. Any software debugger can do this, as those popular ones on Windows can attest.

Of course, there is always give and take. The downsides of software breakpoints are also conspicuous:

- Since the instruction at the breakpoint address has to be replaced, software breakpoints cannot be applied on ROM-based code, which could be a problem for many XIP bootloaders. Although some ICE vendors have devised ways to circumvent this by copying the ROM code to RAM, such an approach would mask out many behaviors unique to the ROM-based code.

- Debugging ISR might also be a problem for software breakpoints, especially in the case when software interrupt is masked out.

- Using SWI is an intrusive way to debug, which could cause conflict with some embedded OS. You should be careful to make sure the SWI numbers (the parameter to the SWI instruction) used by embedded OS do not overlap with those used by ICE.

SIDE NOTES

When I was working for a big telecom firm, I had the opportunity to write code for a SoC chip targeted to cell phone applications. I had a RealView ICE talking to an ARM926 core, with a small real-time OS on it. The OS was provided by a third-party company. However, when I downloaded the image through ICE, it always took a long time (about five seconds) to reach the `main()` function, and the ISR behaved weirdly, which made some threads response time off the mark (very annoying!). Fortunately the day was saved by OS vendor's FAE.

It turns out that the OS was also using SWI, and the SWI number somehow conflicted with RealView ICE's semihosting feature. For those of you who don't know, the semihosting works like this: It has a library on the target side. When users want to print out something, they can use this library to generate SWI and send the data buffer to the host PC through JTAG. Users can thus make a console port out of JTAG. Unfortunately, I was caught up in the SWI index conflict without knowing it.

Like many engineering puzzles you'll face, the remedy seemed easy in retrospect (unchecking the semihosting feature on the ICE GUI in this case), but it took some time to figure it out.

Hardware Breakpoints

To overcome the shortcomings inherited from software breakpoints, CPU designers began to place bus monitoring hardware inside CPU that oversaw the bus activity. Users configure this hardware through JTAG to halt CPU activities when certain memory addresses are being accessed. Since it is all assisted by hardware, hardware breakpoints have no side effects on code execution. In other words, hardware breakpoints are non-intrusive as opposed to its software counterpart, so ROM-based code is no longer unreachable. Unfortunately, the physical constraint of bus monitoring hardware always limits the number of hardware breakpoints available concurrently.

Trace

With JTAG-based ICE and software/hardware breakpoints in the arsenal, engineers can pretty much zap any bugs flying in their faces. However, anyone who stays in this bug-hunting business long enough knows that the most elusive bugs are those that show up intermittently. Sometimes it takes a long while before the error gets triggered.

So when the error does get triggered, it might be nice to have a history of previous CPU activities, which is something ICE cannot offer.

Now imagine this: What if you could have a logic analyzer with an extremely high sample rate and an unlimited sample buffer? Assume you could use this super logic analyzer to probe the CPU instruction bus and data bus. If the error trigger condition were activated, you would know all the instructions that had just been executed before the error happened, which could help you catch the bug a lot easier.

Unfortunately, we don't have such a super logic analyzer right now. Even if we did, no CPU would allow its internal bus to be probed by external instruments. However, what if we built a logic analyzer inside the CPU? Building a logic analyzer inside the CPU is viable and there is a technical term for this: *Trace*.

As shown in Figure 3-7, to implement Trace, the CPU designer would place a capture unit (the specialized logic analyzer) inside the CPU to record all the CPU's activities. The captured activity data is saved in a Trace buffer and later sent out to a Trace box through the Trace port. To save on CPU costs, designer might sometimes move the Trace buffer into the Trace box, and the captured activity data would be compressed and transferred to the Trace box directly with a high-throughput Trace port.

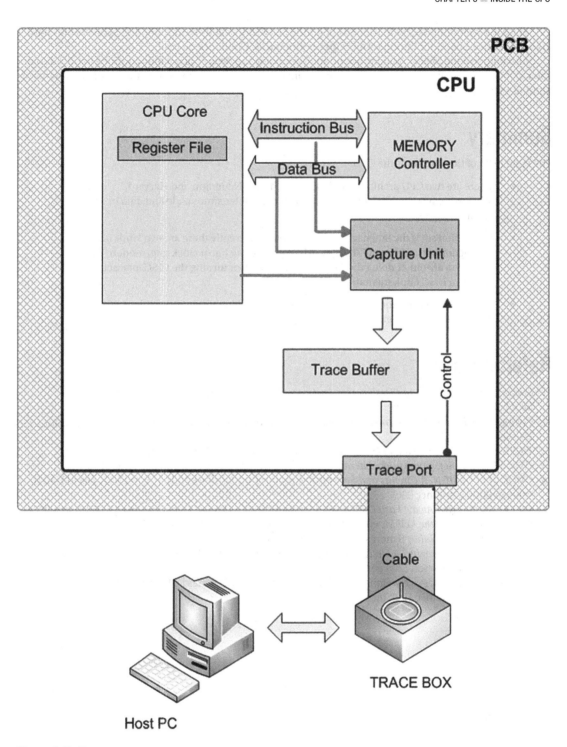

Figure 3-7. *Trace*

Trace is a nice complement to ICE, except for the fact that it usually costs more. So engineers may have to convince their bosses for the necessity of these extra bucks. Good luck on that!

Also note that the idea of "placing a logic analyzer inside the device" is not only confined to CPU design. Later in the book, you'll learn that the FPGA design includes similar ways to capture the history and status of internal logic.

Summary

The main topic of this chapter is the CPU:

- There are two CPU architectures in this world: Von Neumann and Harvard. The former has a unified address space while the latter stores code and data in separate places.

- Instruction set is the language that CPU speaks. Currently there are two kinds of instruction sets: CISC and RISC. By using pipelines to boost clock rate, modern CPU designers are either doing the RISC design directly, or turning the CISC instruction set into a RISC implementation.

- Various CPU debug mechanisms were also explored in this chapter, such as ROM Monitor, ICE, Trace, etc.

References

1. *Computer Organization and Design, the Hardware/Software Interface.* David A. Patterson, John L. Hennessy, Elsevier Inc., 2005
2. *Computer Architecture, A Quantitative Approach, 3rd Edition.* John L. Hennessy, David A. Patterson, Elsevier Science (USA), 2003
3. *MCS 51 Microcontroller Family User's Manual.* Intel Corporation, February, 1994
4. *C55x, V3.x CPU Reference Guide, Literature Num SWPU073E.* Texas Instruments, June, 2009
5. *CMOS VLSI DESIGN, A Circuits and Systems Perspective (3rd Edition).* NEIL H.E. WESTE, David Harris, Pearson Education, Inc., 2005
6. *8051 Keeps Plugging Away: Faster CPUs, Smaller Packages.* Ray Weiss, Electronic Design, February, 2001
7. *Installing and Using the Keil Monitor-51,* Application Note 152. Keil Software, July, 2002
8. *DFT Compiler User Guide.* Synopsys, Inc. December, 2008
9. *Digital Integrated Circuits - A Design Perspective (2nd Edition).* Jan M. Rabaey, Anantha Chandrakasan, Borivoje Nikolic, Pearson Education, Inc., 2003
10. *Silicon VLSI Technology - Fundamentals, Practice and Modeling.* James D. Plummer, Michael D. Deal, Peter B. Griffin, Prentice Hall, Inc., 2000

CHAPTER 4

RAM, DMA, and Interrupt

Yeah, Jesus tried to save me, but there was no space left on his memory card.

—Computer Joke found online by arucardegungrave

Most embedded systems have some sort of memory that can be accessed randomly. There are many types of RAM (random access memory) in this world. Each has a different interface and electrical characteristics. RAM is the main place to store code and exchange data with the CPU core. The throughput of memory has a serious impact on the overall performance. Meanwhile, the use of virtual address vs. physical address in modern OS has made things more complicated for memory access. This chapter serves to map out all these technical details.

Types of Random Access Memory

RAM can take on many forms, with different density, latency, and throughput characteristics.

NVRAM (Non-Volatile Random Access Memory)

This book defines NVRAM as ROM that can be accessed randomly. (There are devices that are actually SRAM plus battery and call themselves NVRAM as well. This book is not trying to be pedantic.) Other than the read-only attribute, NVRAM is treated pretty much the same as RAM, which is mapped into the whole address space.

There are many semi-conductor devices that could be used as NVRAM, like EPROM, EEPROM, and Flash. Most embedded systems developed today use NOR Flash to implement NVRAM.

Traditional SRAM: Intel 8080 vs. Motorola 6800

SRAM typically has a r/w interface like shown in Table 4-1, Figure 4-1, and Figure 4-2. It is called static RAM because unlike Dynamic RAM, it does not need a circuit to refresh the memory cell. Traditional SRAM sometimes is also called 6T SRAM since each SRAM cell is composed of six transistors (Ref [1]). If necessary, the interface can also be made as asynchronous by forgoing the clock signal.

© Changyi Gu 2016
C. Gu, *Building Embedded Systems*, DOI 10.1007/978-1-4842-1919-5_4

Table 4-1. *Typical SRAM Pins*

Signal	Description
CLK	Clock
CS* (or CE)	Chip select (or chip enable)
A	Address bus
OE* (or OD)	Output enable (active low) or output disable (active high)
R/W*	High for read, low for write
D	Data bus

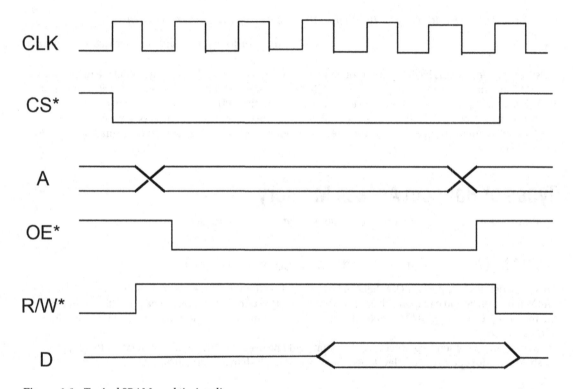

Figure 4-1. *Typical SRAM read timing diagram*

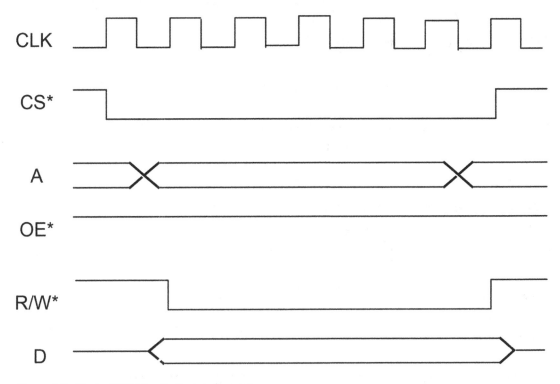

Figure 4-2. *Typical SRAM write timing diagram*

As will be discussed in later chapters, the OE* pin is necessary since the D (Data Bus) is bi-directional. You need this OE pin to control the tri-state output buffer so that the SRAM data output will be in high-impedance during SRAM write. Sometimes the bi-directional D (Data Bus) is replaced by two uni-directional buses (D for Data Input Bus and Q for Data Output Bus), and you don't need OE pin in that case.

Traditional SRAM's latency is pretty low. However, its 6T memory cell limits its density, which translates to high cost. So for RAMs with large sizes, it is not a cost-effective choice.

SRAMs are widely used as external memory for microprocessors. In the 1970s when microprocessors were still considered somewhat state of the art, the two major players in the field were Intel and Motorola. They each released 8080 (Ref [11]) and 6800 (Ref [12][13]) microprocessors into the market. The way that those two microprocessors talk to the SRAM is largely the same as what's mentioned here, with a slight difference in R/W control.

Intel 8080 uses $\acute{R}D$ and $\acute{W}R$ as read and write control, which can correspond to the OE* and R/W* pins in Table 4-1. The drawback of using $\acute{R}D$ and $\acute{W}R$ is that if the microprocessor accidently sets both of them to low, the result will be undetermined. To avoid such ambiguity, Motorola 6800 takes a different approach. It uses two signals called $R/\overline{W}$ and E. The $R/\overline{W}$ is the same as R/W* pin in Table 4-1 functionally, but it only takes effect when E is high. Table 4-2 shows the SRAM signal mapping for Intel 8080 and Motorola 6800 interfaces. Table 4-3 shows the Truth Table for them.

Ostensibly the E signal serves the same purpose as $\overline{CS}$ signal for enabling control. However, the $\overline{CS}$ signal provides enabling for the whole chip, and it can be used to put the SRAM into power saving mode, while E only deals with read and write. In order for the SRAM to work properly, the $\overline{CS}$ signal has to be asserted much earlier as suggested by datasheet to give the SRAM enough time to wake up, but this head start does not apply to the E signal.

Table 4-2. *Signal Mapping for Intel 8080 and Motorola 6800*

SRAM	Intel 8080	Motorola 6800
CLK	*CLK*	*CLK*
CS*	$\overline{CS}$	$\overline{CS}$
A	*A*	*A*
OE*	$\overline{RD}$	$\left(not\,E \right) \vee \left(not\,R / \overline{W} \right)$
R/W*	$\overline{WR}$	$\left(not\,E \right) \vee R / \overline{W}$
D	*D*	*D*

Table 4-3. *Truth Table for Intel 8080 and Motorola 6800*

Function	Intel 8080		Motorola 6800	
	$\overline{RD}$	$\overline{WR}$	E	$R / \overline{W}$
Read	0	1	1	1
Write	1	0	1	0
No Operation	1	1	0	X
Undetermined	0	0		

Although both microprocessors are water under the bridge now, their legacy remains. For example, many LCD modules still offer parallel bus support, which could be either Intel 8080 or Motorola 6800 type, or maybe both. You need to configure the interface correctly before you can send data to it (Ref [14]).

DDR SDRAM (Dual Date Rate Synchronous Dynamic RAM)

Instead of using transistors, Dynamic RAM uses capacitors to build its memory cells. With the improvement of manufacturing technology, Dynamic RAM could achieve a much higher memory density than traditional SRAM. Unfortunately, the high density comes with a necessary evil. Due to the leakage of capacitors, Dynamic RAM cells need to be recharged (refreshed) periodically.

The high density enables DRAM vendors to offer memory chips with larger sizes at lower cost and larger size calls for higher throughput interface. Today all SDRAM chips on the market use dual data rate to transfer data, which means that signals are sampled on both the rising and falling edges of the clock. (There used to be DRAMs with a single data rate interface. They have all faded into oblivion.)

The large size subsequently demands a wider address bus if normal addressing scheme is used. To save address pins, DRAM usually breaks the memory addresses into bank, row, and column, as shown in Figure 4-3. Banks are indexed with dedicated bank address pins, like BA0, BA1. Row and column addresses are multiplexed into one address bus (usually designated with A[n-1 ... 0]).

In Figure 4-3, each small square stands for a memory word. To index a certain memory word, the corresponding row in that bank has to be activated. After the row is activated, those words in that active row can be read or written in burst mode by giving the starting column address.

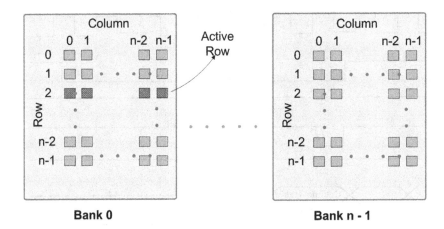

Figure 4-3. *Bank, row, and column addresses*

The technology of DDR SDRAM keeps evolving. The progresses made over generations are mainly in burst length, clock frequency, memory density, and power consumption. However, the data rate still remains twice the clock rate (Ref [2]).

DDR signal pins can be roughly summarized in Table 4-4.

Table 4-4. *DDR SDRAM Signal Pins*

Signal	Description
CK and CK#	Differential clock
CKE	Clock enable
CS#	Chip select
RAS#, CAS#	Row address select and column address select
WE#	Write enable
DM/UDM/LDM	Data mask for write, DM for x32 or wider bus, UDM (upper-byte DM), and LDM (Lower-byte DM) for x16 bus
BA	Bank address
A	Address bus
DQS/UDQS/LDQS	Data strobe; they could be differential signal as well, like DQS/DQSn
DQ	Data bus

The timing diagrams for DDR burst read/write are exemplified in Figure 4-4 and Figure 4-5. As mentioned, the bank/row that contains the target word has to be activated before any read/write can proceed, which corresponds to the bank/row address in Figure 4-4 and Figure 4-5. Be advised that due to the plethora of product offerings, the actual timing diagrams may differ from what is shown here. Always use the manufacturers latest datasheets as your reference point.

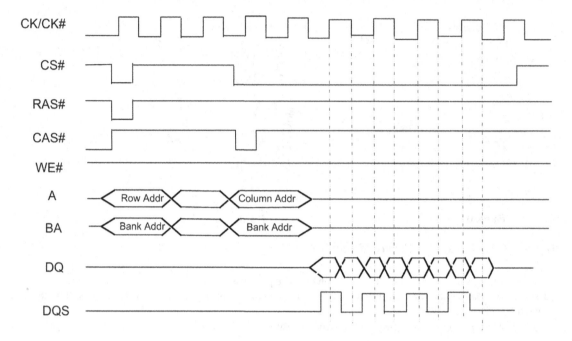

Figure 4-4. *Example of DDR burst read*

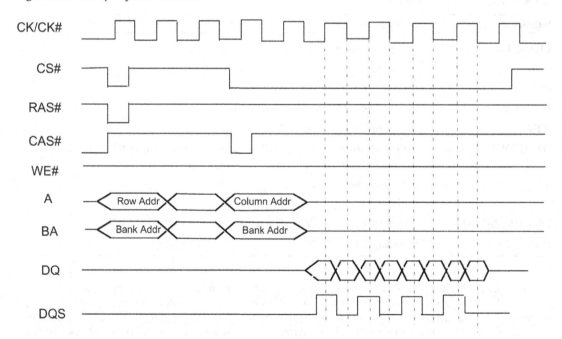

Figure 4-5. *Example of DDR burst write*

Because of the extra refreshing circuit as well as the high throughput, a dedicated DDR controller (DDR PHY) is needed to talk to the DDR SDRAM chip. For the FPGA case, the pin assignment for DDR PHY has to be verified by FPGA software (place and route) before board layout can be started. The reasons are the following:

- To relieve the skew experienced by the parallel bus, DDR interface divides the DQ bus into small groups, with each group taking its associated DQS as the reference clock. The group width can be x4, x8/x9, x16/x18, or x32/x36. If the width is x9, x18, or x36, the group usually also includes the corresponding DM signal for each byte.

- For each I/O bank of the FPGA, the physical pins are also divided into groups in order to implement different DQ group width. The pin assignment has to be well aligned to such physical arrangement. Otherwise the PAR (place and route) will fail due to the physical restriction imposed by FPGA device. In addition, there are other concerns that have to be addressed, like the voltage of the I/O bank, the PLL that is being used, etc. So it is crucial to have FPGA software do a sanity check on the pin assignment before the start of board layout. The time spent on such a check is the time well spent, both electrically and financially.

QDR SRAM[1]

As you saw in the previous sections, DDR SDRAM has achieved a high-throughput interface by transferring data on both edges of the clock. SRAM manufacturer promptly borrowed this page from DDR SDRAM's playbook. Instead of using bi-directional data bus, SRAM manufacturers have divided read and write into separate ports, with each port having its own clock.[2] (However, the address bus is still shared by these two ports.) In other words, SRAM manufacturers have defined something kind of like a dual port RAM interface, with one port for read and the other for write. Both ports can operate at a double date rate. And they call such devices QDR SRAM (Quad Data Rate SRAM), which is, in my opinion, more market propaganda than a real technical term.

The signal pins of QDR SRAM are listed in Table 4-5. As you can imagine, those signal pins in Table 4-5 bear a striking similarity to the ones in Table 4-4, and such similarity is summarized in Table 4-6.

Table 4-5. *QDR SRAM Signal Pins*

Signal	Description
K and K#	Differential clock for write.
C and C#	Differential clock for read if single clock mode is not used. (Exists on QDR II only; recommended when multiple QDR II devices are used. See Ref [4][5] for more detail.)
CQ and CQ#	Differential clock as echo clock. Can be used as a read clock.
WPS#	Write port select.
RPS#	Read port select.
BWS#	Byte write select.
A	Address bus.
D	Data bus for write.
Q	Data bus for read.
QVLD	Read data valid.

---

[1]This book uses QDR SRAM as a generic term, which covers all the QDR type SRAM, including QDR II and QDR II+.
[2]Single clock mode is also possible for many QDR SRAM chips.

Table 4-6. *Resemblance of QDR SRAM and DDR SDRAM Signals*

QDR SRAM Signal	Corresponding DDR SDRAM Signal
K and K#	DQS/DQS#
CQ and CQ#	
C and C#	
WPS#	CS#
RPS#	
BWS#	DM
A	A
D	DQ
Q	

However, unlike DDR SDRAM, QDR SRAM does not multiplex addresses into rows and columns, so there is no row-activate command for SRAM. Other than that, the burst read and burst write of QDR SRAM also resemble their counterparts in DDR SDRAM.

Other Types of RAM

In addition to DDR SDRAM and QDR SRAM, there are other types of RAM that are used in embedded system design, like RLDRAM and PSRAM.

- *RLDRAM*

    The interface of RLDRAM (Reduced Latency DRAM) can be either Separate I/O (SIO) or Common I/O (CIO). If read and write ports are separated, the interface is close to QDR SRAM. If the data bus is bi-directional, the interface is very much like DDR SDRAM.

    As its name suggests, RLDRAM has a very low cycle time (tRC[3]). To demonstrate this quantitatively, let's compare the tRC from an 800MHz DDR3 SDRAM (Ref [6]) to that from a 400MHz CIO RLDRAM II (Ref [7]). In fairness, both of them are offered by the same manufacturer (Micron Technology, Inc.). The 800MHz DDR3 gives you a tRC of 50ns (Ref [6]) while that number is only 20ns for the 400MHz RLDRM II (Ref [7]). Of course, this is still no match to the latency of QDR SRAM. (Take the 550MHz QDR II+ in Ref [3] for example' it has 2.5-cycle read latency, which is equivalent to 4.55ns.) However, it does offer a low cost alternative to QDR SRAM.

- *PSRAM*

    As you can see, the need for refreshing the circuit is DRAM's soft belly. To overcome this shortcoming, memory manufacturers started to offer something called PSRAM (Pseudo-SRAM). Conceptually, it can be seen as a DRAM chip that carries the refreshing circuit inside, and outwardly it looks like a SRAM. I will leave its tech details to those inquisitive readers.

---

[3]tRC (Row Cycle Time) for DRAM is defined as the minimal time gap between two successive row active commands in the same bank.

PIO and DMA

Despite the multitude of RAM types, the sole purpose of RAM is to store data. Generally speaking, there are only two ways to access data stored in RAM: PIO and DMA.

PIO

PIO stands for Programmed Input/Output, which means all the memory R/W is managed by the CPU, like the code snippet in Listing 4-1.

Listing 4-1. Use PIO to Access Memory

```
int buffer[256];
int *p;
int i;

p = buffer;

for (i = 0; i < 256; ++i) {

    *p++ = i;

}
```

The plus side of PIO mode is that:

- The coding is very straightforward; there is no interrupt handler involved.
- Everything is done through a virtual address. No need to convert the virtual address into a physical address.

However, the flip side is that:

- It consumes a great deal of CPU time. In other words, it carries a big CPU overhead.
- As you can see in Listing 4-1, the PIO actually breaks memory access into 256 single memory write operations. All the burst modes discussed so far cannot be utilized by PIO. So it carries huge performance penalty.
- After the code in Listing 4-1 is executed, the data might still remain in cache instead of the memory. A cache flush is needed. (Cache and memory consistency will be discussed shortly.)

So PIO is not suitable for transferring data with high throughput. To relieve CPU from those chores, you need the assistance of extra hardware.

DMA and MMU

In order to boost system performance and save the precious CPU cycles, DMA (Direct Memory Access) is introduced. An extra piece of hardware called DMA controller is placed onto the memory bus to off-load the CPU core from the job of large data transferring. The CPU core has to configure the DMA controller and kick off the data transferring. When the job is done, the DMA controller will notify the CPU core through interrupt. And CPU core gets to execute more important jobs in between.

There is no doubt that DMA holds a great performance advantage over PIO. RAM chips would blissfully churn out long bursts for DMA, much like those you saw in Figure 4-4 and Figure 4-5. However, the outsourcing of the data-transferring job from the CPU core to the DMA controller inevitably has its technical consequences.

As we all know, it is software that pulls the strings inside the CPU. One of the primary software players is the OS (Operation System). Most operating systems use virtual addresses instead of the actual physical addresses to manage their memory operations. The use of virtual addresses has its virtues:

- It makes OS less coupled with the actual hardware it is running on, so it becomes easier to port the OS to different hardware platforms.

- The virtual address would fool every task running on the OS into thinking that it is the only task that runs the show. Although such illusion may sound absurd on the first encounter, it does make multi-tasking a lot easier since each task runs in its own virtual address space. However, it also means that two different tasks could operate on the same virtual address without knowing each other.

To facilitate the implementation of a virtual address, CPU usually carries a piece of hardware called MMU (Memory Management Unit). MMU does the translation from virtual address to physical address in the hardware. When the OS schedules a new task to run on the CPU, it will configure the MMU differently so that the same virtual address of each task is mapped to a different physical address. Thus multiple tasks can coexist peacefully in the physical memory space.

To do the address translation more efficiently, memory is divided into pages, with each page having a fixed size. A typical page size is 4KB, although other sizes are also possible. OS would maintain a page table and manage the memory by the unit of pages instead of words or bytes. This implies that *two consecutive pages in the virtual address space could be mapped to two separate pages in the physical address spaces.* That's what complicates the DMA!

When software tries to move data in large sizes, it usually means to read or write a big memory buffer in the virtual address space. Needless to say, such a buffer is continuous by the account of virtual address. However, after the translation through MMU, the big buffer becomes a bunch of pages scattered all over the physical memory space. Since the DMA controller operates independent of CPU core and MMU, the DMA controller has to live with such inconveniences.

To get the job done, the DMA controller usually operates on a data structure called the *scatter-gather list*. Although the exact formats of the scatter-gather lists differs depending on the hardware, they share a lot of common traits with the example shown in Figure 4-6 (assuming a 32-bit address space).

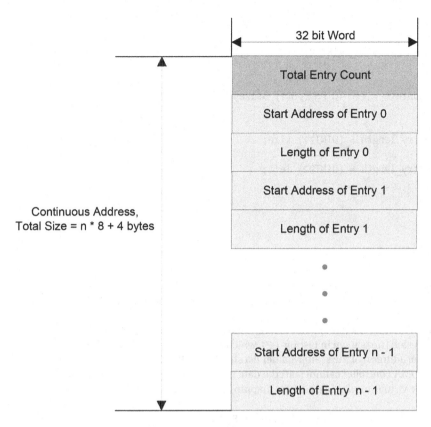

Figure 4-6. *A typical scatter-gather list*

The scatter-gather list shown in Figure 4-6 resides continuously in physical memory space. Most DMA controllers require the scatter-gather list to be stored continuously in physical memory, although formats like link-list type are also possible. In practice, the majority of the memory blocks referenced in Figure 4-6 would start on a page boundary and have a length of page size or the multiple of page size, which is to be expected due to the page table mapping.

Because of the scatter-gather list, the programming of DMA is much more tedious than that of PIO. First of all, as demonstrated in Listing 4-2, the continuous buffer in virtual address space has to be translated into a bunch of scattered physical memory blocks, with each block corresponding to an entry in Figure 4-6.

Listing 4-2. Pseudo-Code to Construct Scatter-Gather List Entries

```
UINT32 *pBuf;
INT32 left, size;

typedef struct {
   UINT32 start_address;
   UINT32 entry_length;
}SG_ENTRY;

SG_ENTRY *pSG_Entry;

pBuf = THE_STARTING_ADDRESS_OF_VIRTUAL_BUFFER;
```

```
left = TOTAL_TRANSFER_SIZE;

pSG_Entry = THE_VIRTUAL_ADDRESS_OF_ENTRY_0;

while(left > 0){
   pSG_Entry -> start_address = VIR_TO_PHY(pBuf, &size);

   if (left > size) {
       pSG_Entry -> entry_length = (UINT32) size;
   } else {
       pSG_Entry -> entry_length = (UINT32) left;

   }
   pBuf += size;
   left -= size;
   ++pSG_Entry;
};
```

The OS would provide a function call to help the mapping from virtual addresses to physical addresses, such as the VIR_TO_PHY call shown in the pseudo-code. This size parameter is the output of VIR_TO_PHY call, which indicated the length of corresponding physical block.

After all the physical blocks have been mapped out, the "total entry count" in Figure 4-6 also needs to be filled with the correct value (not shown in Listing 4-2), which completes the scatter-gather list. Then the starting address (physical address) of the scatter-gather list should be filled into the DMA controller's configuration register. The DMA controller cannot start the data transferring until it has the complete scatter-gather list. The whole virtual-to-physical address mapping is illustrated in Figure 4-7.

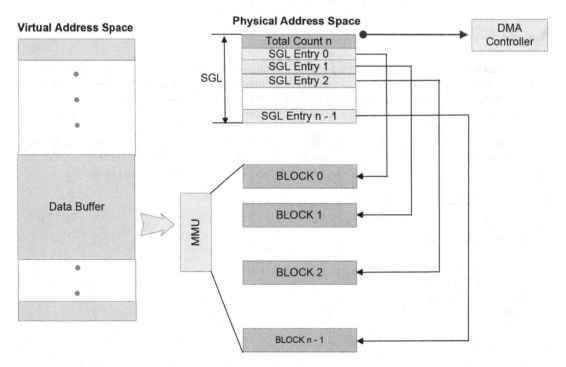

Figure 4-7. *Virtual-to-physical address mapping*

When the data transferring is finished, the DMA controller will notify the CPU core through interrupt. Sometimes in order to make the programming flow less complicated, the software could choose to disable the DMA interrupt during the configuration stage. Instead of waiting for the interrupt, the software keeps polling the DMA status register until the done flag is set, as illustrated in Listing 4-3. In this way, the software does not have to implement the interrupt handler.

Listing 4-3. Polling DMA Status

```
*) To construct the complete Scatter-Gather List

*) set up the DMA controller with the SGL, but disable the DMA interrupt

*) kick off the DMA transferring

do { // polling the DMA status register

    status = read_DMA_status_register();

} while (!(status & DONE_FLAG));

// DMA is finished
```

Since the CPU core is actively involved in the status polling, this type of "DMA without interrupt" practice is still sometimes called PIO, although data transferring is done by the DMA controller with the benefit of burst transfer.

Memory Allocation for DMA

As you know from previous sections, in order to make DMA work, you need to prepare a data structure called SGL (Scatter-Gather List) in memory, which implies that you might have to allocate memory space for SGL.

Those who have ever programmed in C know that you can ask for more memory by calling the malloc() function. Consequently it begs the question: Can you use malloc() to allocate space for SGL? In addition, if you want to allocate scratch buffers for DMA, can you use malloc() to do that job as well?

This section covers my suggestions in this regard.

First of all, C Library functions, like malloc(), are not recommended in this case. Since you are programming the DMA controller, most likely you are either writing a device driver for the OS or working entirely with the bare-bones hardware. *As a general guideline, try to avoid using C Library functions as much as you can when you interact with hardware directly.* This is because the actual implementation of the C Library depends on the OS and the compiler it is based on. Since you don't know for sure what underlying OS API it is using, you run the risk of shooting things in the dark.

And if you have to allocate memory inside an ISR, the allocation function cannot cause blocking. Some allocation functions could cause blocking because they have to wait on pages to be released or swapped back into the memory. If you use OS API, you should read the remarks carefully to make sure you pick a non-blocking function. Due to the ambiguity of its actual implementation, you have no straight way to judge a malloc() function on that.

There are usually two types of memory pool OS manages: paged memory and non-paged memory. The difference between those two is that paged memory can be swapped out of the physical memory space onto a hard disk or Flash as the OS sees fit, while non-paged ones always stay in the memory. One example is the page file used by Windows, which stores images of paged memory that is swapped out to the hard disk by the OS. To make the DMA operation successful, only non-paged memory can be used for SGL and DMA buffers;

otherwise, you could suffer page-fault exceptions. So you need to specify certain parameter flags (like the GFP_DMA flag in Linux) or call designated OS APIs when you ask for memory from OS. This is another reason why I strongly discourage using malloc() here. Chances are that malloc() will dole out paged memory at some point to sabotage your DMA operation.

For embedded systems, if the existing size of the physical memory suffices your needs, you could choose to disable the page swap for the peace of mind. (Oftentimes this is the only option due to the absence of hard disk space or viable Flash space in the system.) In fact you could also choose to manage part of the physical memory on your own. For example, assume you have 512MB RAM in the system. You could configure the embedded OS to let it see only the lower 256MB and use the higher 256MB as DMA buffers, allocating and releasing them in the device driver you are providing.

Cache/Write Buffer and Memory Incoherence

Up to this point, assume you've done everything by the book to set up the DMA, constructing the SGL with the correct type of memory, and configuring the DMA controller meticulously. Is there still anything left that could torpedo your DMA operation? I say, "Watch out for the cache!"

Today almost every CPU carries an internal cache. This cache is made of high-speed SRAM. When the CPU reads something from memory, part of the data will be placed in the cache. The philosophy behind this is called *temporal locality,* which bets that the memory location being accessed now will be accessed again in the near future.

Cache is implemented by hardware, and it is largely transparent to programmers in normal operation. Sometimes you see terms like "x way set associative cache" in the CPU datasheet, and their corresponding definitions can be found in various text books. But practically such terms are only meaningful to the CPU designer. At the system level, cache is just a black box that keeps another copy of the data besides memory.

However, the DMA controller sits on the memory bus and has no knowledge of the cache's internal status. So it is possible that when DMA overwrites the data at a certain address in memory, the cache still keeps another copy of the obsolete data with the same memory address. Such unfortunate occurrence is called *memory incoherence.*

Memory incoherence can also happen in write. Under the principle of "temporal locality," the CPU chooses to keep part of the write data in cache instead of flushing it to memory right away (assuming the cache is set in write back mode). In addition, to merge small writes into bigger trunks, many CPUs also place a write buffer between the cache and memory. Before DMA takes off, the corresponding data in the cache and write buffer have to be flushed into memory to prevent memory incoherence.

On a desktop PC, the x86 CPU performs bus snooping to closely watch the memory bus and compare notes with its internal cache. When it detects R/W activity on the memory bus, it will invalidate or flush its internal cache accordingly if there is a cache hit. But such approach complicates the CPU design and brings added cost. In the world of embedded systems, most CPUs, such as ARM, choose to let software handle this with the intent to save cost. Thus the burden falls on the programmers' shoulders to flush the cache/write buffer before DMA starts, or invalidate the cache for read when DMA has occurred. Fortunately, embedded OS usually provides functions to facilitate such an operation. However, keep in mind that cache operates in the granularity of cache line, whose size is usually bigger than the CPU's word length. For example, a 32-bit CPU might have a cache with a 128-bit cache line size. Thus it is strongly suggested to align the memory buffer to the cache line size boundary during memory allocation.

Interrupt

On the heel of DMA, I think it is a good opportunity to embark on the subject of interrupt, since they usually work hand in hand.

Interrupt Mechanism

Interrupt is adopted almost everywhere in the embedded systems: the CPU core delegates certain jobs to peripherals (external devices). When the job is finished, the peripheral would notify the CPU core by signaling through the interrupt pin.[4] The CPU gets to handle other tasks in the meantime. Interrupt is a good way to increase parallelism of the whole system, which usually means higher efficiency and better performance. As illustrated in the previous sections, DMA is a typical example to demonstrate the benefit of interrupt over polling.

Interrupts can come from multiple sources. They can be from power reset, timer, external devices, software interrupts, etc. In the old days when the scale of the system was small and the number of external devices was limited, the CPU designer would place the entire interrupt controller inside the CPU core and reserve a small number of pins for external interrupts. A typical case is Intel 8051, which has only two interrupt pins (INT0* and INT1*) for external devices.

When an interrupt is generated by its source, and if that interrupt has the highest priority among all the interrupts to be serviced, the CPU will save the current status and then point its PC counter to a fixed address associated with that interrupt. Software should place a branch instruction (or instructions of similar nature) in that address, which jumps to the actual ISR. The whole collection of these fixed addresses and branch instructions is called IVT (Interrupt Vector Table), because they are usually stored consecutively in a pre-determined order, with one address/branch instruction per each interrupt. As illustrated in Figure 4-8, IVT is traditionally located at the top or bottom of the whole address space, with power reset as the first vector in that table. Each interrupt pin reserved for external devices will have a corresponding entry in the IVT.

With the explosive growth of both market and technology, the scale of embedded systems today is much larger than in the past. And the number of external devices has also ballooned accordingly, which puts a higher demand on interrupt pins. One thing that can be done to solve this supply/demand issue is, of course, to increase the number of interrupt pins available. However, such approach requires the augment of IVT size and the redesign of CPU core, which is costly and unfeasible in practice. Fortunately, engineers have found other ways to balance the supply/demand. One way is to share those interrupts; the other way is to use a Vectored Interrupt Controller.

---

[4]Interrupt can also be signaled by sending messages instead of using dedicated pins. Such interrupt mechanism is called MSI. MSI is mainly used on the PCI local bus (optional, Ref [19]) and PCI express bus (Mandatory, Ref [18]). MSI is explored in later sections.

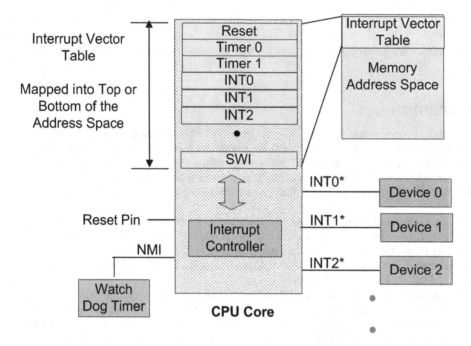

Figure 4-8. *Interrupt and IVT*

Shared Interrupt

If the supply of available interrupt pins is limited while the demand remains high, you could make do by letting multiple external devices share the same interrupt pin and checking the device register inside ISR to determine the interrupt source.

However, to reduce the risk of in-fighting, the electrical characteristics of those devices should be checked out before you let them share an interrupt. There are two main aspects on that score, discussed next.

Edge Triggered Interrupt and Level Triggered Interrupt

Electrically speaking, there are two types of interrupts: *edge triggered* and *level triggered*. For edge triggered interrupts, the device will send out a pulse to the CPU, and the CPU will pick up the rising or falling edge of that signal by using edge detection or direct sampling. On the other hand, the device of a level triggered interrupt will set the signal level to high or low (most likely it will be active low) and keep that level until it gets acknowledgement from CPU.

The problem with the edge trigger interrupt is that:

- The CPU needs to have hardware to keep track of those pulses. Otherwise, low-priority interrupt may be missed if the CPU is servicing high-priority interrupts.

- The sharing of edge triggered interrupt is also a challenge. If two pulses from different devices come at close range, such as one happens in the middle of the other, those two pulses will be superimposed on the shared interrupt pin as one longer pulse, which means one of the interrupt pulses could slip through the crack.

These shortcomings make edge triggered interrupt less desirable to many system designers.[5] In fact, with the improvement response time of external devices, level triggered interrupt becomes much more popular for out-of-band interrupt signaling. The thing with level trigger devices is that the device should de-assert its interrupt pin in a short period of time after the CPU acknowledges the interrupt; otherwise, the CPU may be lulled into taking the same interrupt twice.

Push-Pull and Open Drain (or Open Collector)

Sharing interrupts basically means merging the interrupt pins of two (or more) devices. Since the interrupt pins are usually the output of these devices, a close inspection of their electrical spec is necessary. For CMOS devices, their output is push-pull or open drain.[6]

1. *Push-Pull*[7]

   For CMOS circuits, they are usually composed of two parts—pull-up network and pull-down network—which are made up of PMOS and NMOS, respectively [1]. And at the output stage, the I/O port is called push-pull if both the pull-up and pull-down are present. A simple case is the output of an inverter, which is shown in Figure 4-9.[8]

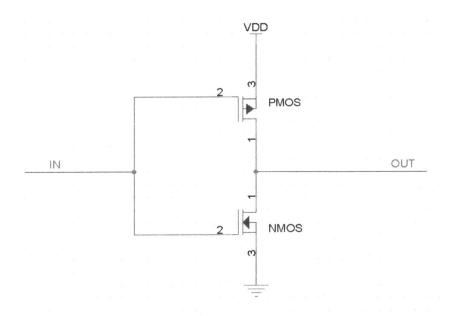

Figure 4-9. An example of push-pull output

---

[5]This only holds true for out-of-band interrupt signaling. With the advance in serial link technology and point-to-point interconnection/switch technology, edge triggered interrupt is actually favored for in-band interrupt signaling. In-band interrupt signaling, mainly MSI, is discussed in later sections.

[6]Theoretically there is a third option that is called *line driver*. Line driver is the very opposite of OD/OC, but it is not as common as OD/OC for the I/O port design.

[7]Its bipolar counterpart is sometimes called *totem pole*.

[8]It is practically the same for bipolar circuits. Just replace the NMOS and PMOS with NPN and PNP transistors.

When a push-pull port wants to output a logic 1, it will raise its output voltage to VDD. So when two push-pull outputs from different devices share the same interrupt, they cannot be connected directly. Instead they should go through an AND or OR gate, depending on the polarity of the output, as illustrated in Figure 4-10.

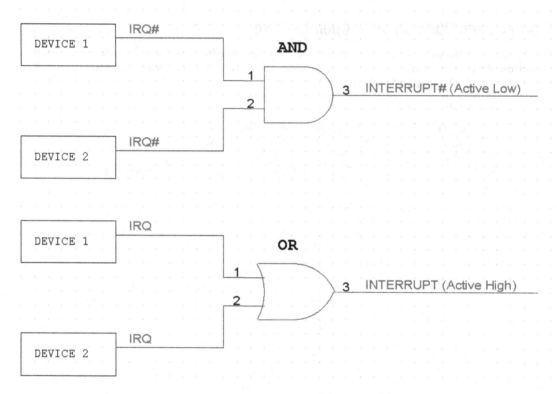

Figure 4-10. *Share interrupt between push-pull output*

However, in a real-world design, an embedded system is actually a medley of components from various sources. To reduce cost, sometimes the procurement and logistic people are involved in the parts-picking as well. Thus, it is highly likely that more than one power rail is needed. For example, some legacy components may require 3.3V supply while newer components only need 1.8V I/O. For push-pull drivers, this could cause headaches for the designer if devices sharing the same interrupt require different VDD/VCC levels or the AND/OR gate has a different voltage level from those devices. Although you can always throw voltage shifters in between, it will inevitably lead to higher BOM cost. And this is the main reason why so many devices make their interrupt output as open drain instead of push-pull.

2. *OD (Open Drain[9])*

Different from push-pull, open drain output does not have those pull-up PMOS. When an OD circuit tries to output a logic 1, it actually sets the output as high impedance. To use the OD output, an external pull-up resistor is needed, as shown in Figure 4-11. The pull-up resistor should be big enough (such as 10KΩ) to keep the current below the allowable sink current. Since the output is high impedance, you can connect two OD output directly without going through extra AND/OR logic gate, and the VDD used to pull up the output does not have to be the same as is used to power the device. Such characteristics also imply that the open drain interrupt is always active low.

Open drain is widely used in interrupt sharing. However, the weak (big value) pull-up resistor could introduce a big RC constant, which may limit the toggle rate of the interrupt signal. This should be taken into consideration as well when you choose the value of the pull-up resistor and what I/O devices you want to use/share.

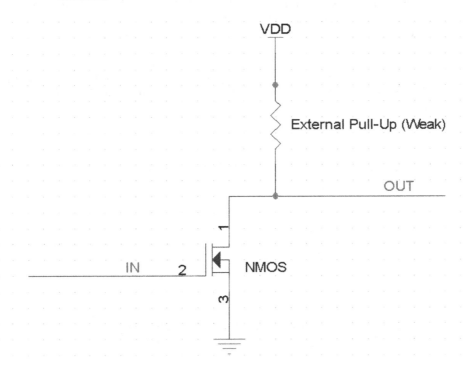

Figure 4-11. *Open drain with weak pull-up*

---

[9]Its counterpart in a bipolar circuit is called *open collector*.

VIC (Vectored Interrupt Controller)

With the increasing demand for interrupt pins, CPU designers soon realized that they needed to strip the interrupt controller off their "core business". The external interrupt controller can be replaced or upgraded later to keep up with the I/O demand, while the CPU core stays the same. In this way, they can have a CPU core with a relatively long shelf life while enjoying the flexibility brought by an external interrupt controller. To fulfill this vision, the interface between the CPU core and the interrupt controller has to be consistent across the board, thus the Vectored Interrupt Controller is introduced.

As shown in Figure 4-12, VIC has a bunch of INT*/IRQ pins that can be attached to external devices. For each INT*/IRQ pin, its priority and masking can be configured through control registers. To expedite interrupt handling, some VICs let users set up the corresponding ISR address (Interrupt Vector) for each INT*/IRQ pin as well. When interrupt happens, the unmasked INT*/IRQ with the highest priority will be picked out, and its ISR address or INT*/IRQ number will be placed in a register. Meanwhile, VIC will assert the INT*/IRQ pin on CPU core to notify the coming interrupt.

When receiving the interrupt notification, if the CPU core is not doing anything of higher priority, it will save the CPU status and jump to the INT*/IRQ handler, which is a generic handler for all the interrupts coming from VIC. Inside that generic handler, the software would read out the VIC register to determine the ISR address or INT number for the incoming interrupt. If ISR address is provided by VIC, the generic handler would jump to that address directly; otherwise, the INT number will be used as an index to determine the ISR entry address for that particular INT*/IRQ number.

As shown in Figure 4-12, a standalone VIC external to the CPU core brings flexibility. A typical VIC can support 32-64 external devices simultaneously. If more interrupt pins are needed, you can either replace the VIC with one that has bigger capacity or use daisy chain to get more interrupt pins. And SWI implement is also made easy by using external VIC.

As shown in

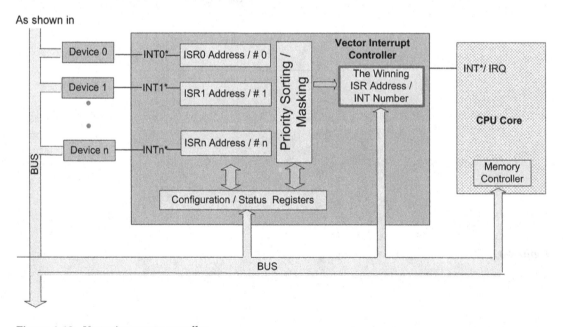

Figure 4-12. *Vector interrupt controller*

Interrupt from Watchdog Timer

Every human endeavor is accompanied by mistakes. Embedded systems are certainly far from bulletproof. Things like process variation, voltage drop, temperature fluctuation, etc. could easily throw a wrench in the hardware. Meanwhile, plagued by cunning bugs, software is no better in this regard. I bet you must have seen a man-made natural phenomenon called "blue screen" at least once in your lifetime. :-)

The bottom line is, in spite of our due diligence, we have to face the possibility of having hidden defects inside our products. However, when things do go awry, we don't want the system to become a basket case. Instead, we hope the system could jump to a fail-safe mode in a timely fashion, and the system should take all the proper measures in that fail-safe mode for recovery, like cutting off the high voltage power, turning off the engine, or simply resetting the whole system. Hopefully if the hidden defects can only be triggered by a peculiar confluence of unfortunate events, the system will be restored to normal mode and continue to function for a long while. The total down time is reduced to a minimum in that sense.

Such a fail-safe idea is fulfilled by a device called the *watchdog timer*. Basically, the watchdog timer is a counter that counts down from a predetermined value. As its name suggests, the device works like a pettish dog. We need to pat the dog from time to time to please it. (Reload the counter value before it reaches zero.) If we neglect the dog for too long, the dog will be mad and start barking. (The counter reaches zero, which triggers the fail-safe.) Hence, a common usage of a watchdog timer is to tie its output to the CPU core's NMI pin and let the software handle it as the top-priority interrupt. Software can even log debug status before it resets the system. But if safety and reliability are big concerns, the watchdog timer output can also be hardwired to something like a power switch or valve control without software intervention.

The watchdog timer scheme seems artless in a single task setting when the main code is merely a big loop with a few ISRs for interrupt handling. However, it will become much more involved when you try to reload the watchdog timer on a RTOS with multi-task running. Thorny problems like dead-lock or live-lock will soon arise. Fortunately, a comprehensive treatment of this topic can be found in Ref [8]. I strongly suggest readers peruse this article before writing any code for a watchdog timer.

Message Signaled Interrupt (MSI)

For all the interrupt mechanisms you have discussed so far, each device must have a dedicated interrupt pin external to the data bus. Such interrupt mechanism is called an *out-of-band interrupt*. An out-of-band interrupt enjoys its straightforwardness. However, as discussed in previous sections, when the system scales up, the increasing number of interrupt pins makes the whole system bulky and unwieldy.

Fortunately, with the advancement in serial link technology and switch technology, point-to-point topology becomes popular in large-scale systems. PCI express is a typical representation of such system (Ref [10]). In PCI express, in-band interrupt mechanism is both favored and required[10]. Instead of using separate interrupt pins, devices signal the interrupt by sending special message on the data bus, writing to certain memory locations. Therefore, it is also called MSI (Message Signaled Interrupt). The plus side of MSI is that there is no distinctive difference between the data and the interrupt, so the implementation can be consistent across the board. The system can thus be scaled up fairly easily. However, MSI is more complicated than its out-of-band counterpart and requires software support as well. (For example, on Microsoft Windows, only Windows Vista or higher version support MSI.) And instead of using level-trigger, MSI is actually edge-triggered interrupt mechanism (Ref [10]) due to its message based nature. So for small- and middle-scale embedded system, out-of-band interrupt is still largely favored for its simplicity.

---

[10]In-band interrupt (MSI) is optional in PCI local bus 2.2 and 3.0.

Summary

This chapter is mainly about memory access and interrupt:

- There are various RAM types that you can choose from to achieve random access. But there are only two ways to do the read and write: PIO and DMA. DMA is able to fully take advantage of the burst mode that most RAM supports, and it often works hand in hand with interrupts.

- Interrupts can be edge triggered or level triggered. Interrupt pins can be shared, or there could be a dedicated VIC to expand the number of IRQ pins.

References

1. *CMOS VLSI DESIGN, A Circuits and Systems Perspective (3rd Edition)*. NEIL H.E. WESTE, DAVID HARRIS, Pearson Education, Inc., 2005
2. *External Memory Interface Handbook*. Altera Corporation, 2010
3. *CY7C1561KV18, CY7C1576KV18, CY7C1565KV18 72-Mbit QDR II+ SRAM 4-Word Burst Architecture (2.5 Cycle Read Latency)*. Cypress Semiconductor Corporation, July, 2010
4. *CY7C1510KV18, CY7C1525KV18, CY7C1512KV18, CY7C1514KV18 72-Mbit QDR II SRAM 2-Word Burst Architecture*. Cypress Semiconductor Corporation, April, 2011
5. *QDR™-II, QDR-II+, DDR-II, and DDR-II+ Design Guide,* AN4065, Cypress Semiconductor Corporation, November, 2007
6. DDR3 SDRAM (MT41J256M4 – 32 Meg x 4 x 8 banks, MT41J128M8 – 16 Meg x 8 x 8 banks, MT41J64M16 – 8 Meg x 16 x 8 banks) Datasheet Rev I, Micron Technology, Inc., February, 2010
7. CIO RLDRAM II (MT49H32M9 – 32 Meg x 9 x 8 Banks, MT49H16M18 – 16 Meg x 18 x 8 Banks, MT49H8M36 – 8 Meg x 36 x 8 Banks) Datasheet Rev N, Micron Technology, Inc., May 2008
8. "Watchdog Timers," by Niall Murphy, *Embedded System Programming,* November, 2000
9. *LVDS Owner's Manual (4th Edition)*. National Semiconductor, 2008
10. "Best Practices for the Quartus II TimeQuest Timing Analyzer," (Chapter 7, *Quartus II Handbook,* Version 10.1, Vol 3)
11. *Intel 8080 Microcomputer Systems User's Manual,* September, 1975
12. MC6800 8-bit Microprocessing Unit (MPU), Motorola Semiconductor Products Inc., 1984
13. *M6800 Programming Reference Manual 1st Edition,* Motorola, Inc., 1976
14. AN10880, Using the LCD interface in the LPC313x/LPC314x/LPC315x devices, Rev.01, NXP Semiconductors, October, 2009

CHAPTER 5

Bus Architecture

All parts should go together without forcing. You must remember that the parts you are reassembling were disassembled by you. Therefore, if you can't get them together again, there must be a reason. By all means, do not use a hammer.

—IBM Maintenance Manual, 1925

Inside an embedded system are numerous components, such as the CPU, memory, the disk controller, etc. Inevitably these components need to exchange data, and data is moved from one component to another through interconnections. We call these interconnections the *bus*. There are many bus architectures existing in the world of embedded systems. This chapter covers some crucial aspects of them. Popular bus standards like PCI, USB, RS-232, I2C, and SPI will be discussed.

In the past decade, serial bus has gradually unseated parallel bus for the sake of higher throughput. This may sound counterintuitive on the first encounter, and the reasons will be revealed later in this chapter as well.

With those goals in mind, let's start our "bus ride".

Bus with System Synchronous Timing

When we were learning digital design in school, we were told that synchronous design is the way to go. Part of this doctrine is that we should stick with single clock domain as much as we can, although sometimes clock domain crossing or asynchronous input is inevitable.[1] Such doctrine has also left its fingerprint on bus architectures. If all the devices on the bus use one global clock source to transfer data, we call such bus clocking scheme *system synchronous timing*, as shown in Figure 5-1.

The well-known PCI local bus (Conventional Parallel PCI) is one of the adopters for system synchronous timing. A global clock would no doubt simplify the bus protocol, but its drawbacks are also readily apparent:

- To reduce clock skew, the clock traces have to be matched for length. If expansion card is used, the trace length on the expansion card has to be taken into account as well. With the number of devices increasing, this actually puts on a strain on board layout.

- Due to tolerance in the fabrication process, we can expect variations among different areas of the PCB, which further aggravates clock uncertainty.

---

[1]Clock domain crossing and handling asynchronous input are discussed in later chapters.

© Changyi Gu 2016
C. Gu, *Building Embedded Systems*, DOI 10.1007/978-1-4842-1919-5_5

- The drive strength of the clock buffer also limits the maximum number of loads as well as the maximum clock rate on the bus.

- Due to the mechanical constraints, the system synchronous timing is sometimes virtually impossible. For example, some externally attached devices require a long cable to connect to the bus, and that would introduce too much clock skew with system synchronous timing.

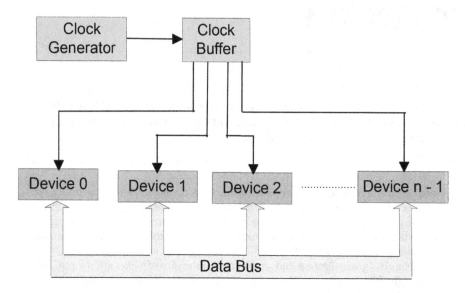

Figure 5-1. *System synchronous timing*

With the increasing demand for more devices and higher bus speed, it doesn't take long for the epiphany to come: Why don't we send the clock along with the data?

Bus with Source Synchronous Timing

As shown in Figure 5-2, under source synchronous timing, device A would send the clock along with the data to the device B. For now, assume the data bus is parallel bus with single ended signal.[2] The bus can be either uni-directional or bi-directional. (If it is bi-directional, you need the R/W signal in Figure 5-2 to tell you the direction.) The topology can be point-to-point, where one master sends out the clock/data to one slave, as is the case in Figure 5-2. Or it could be one-to-many, with one master and multiple slaves, and data can only be transferred between the master and the active slave (chosen by the chip-select signal). One such example is the GPMC bus in Texas Instruments OMAP processor, which supports one-to-many topology with bi-directional data bus.

---

[2]A serial bus with a differential signal is discussed in later sections.

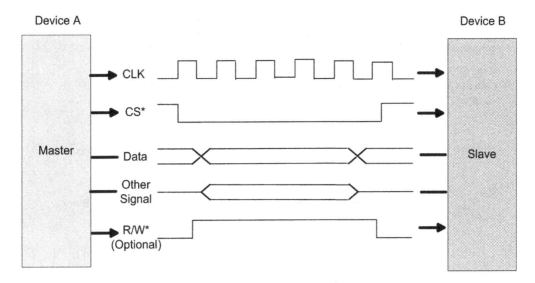

Figure 5-2. *Source synchronous timing*

The good news is that now only the trace length between individual devices needs to be matched. (As opposed to the system synchronous timing, where the clock trace is global.) Since the clock goes pretty much the same routing as data bus, hopefully the fabrication process and temperature fluctuation would leave the same amount of deviation on both the data bus and clock. In this way, clock uncertainty can be reduced while the layout can be more flexible.

Oftentimes, devices can be implemented in FPGA. Should you choose to implement device A or B (or both) in FPGA, here are a few caveats in addition to the guidelines for PCB layout:

- As illustrated in Figure 5-3, for each IO pin on the FPGA, the FPGA would have a corresponding IO cell, which has registers inside. A parallel data bus must have multiple data lines. To reduce the uncertainty among those data lines, it is better to register them all in the IO cells. Otherwise, those data lines would have to be registered inside the FPGA fabric, in which case each data signal may experience different delay due to the routing difference. And those routing delay may constantly change from one synthesis iteration to another. Thus extra uncertainty might be introduced if they are registered inside FPGA fabric instead of the IO cells. (For Altera FPGA, this means the Fast Input Register and Fast Output Register options have to be enabled in the assignment editor.)

- To alleviate ringing and reflection on the bus, the output strength of data bus should also be carefully controlled. (For Altera FPGA, this means selecting the proper Current Strength value in assignment editor.)

- Also, the entire design should be carefully constrained for timing, which largely means finding out the proper input and output delay according to bus spec. (For Altera FPGA, more information can be found in Ref [2].)

- For better jittery handling, you can use PLL to track the bus clock.[3] If you choose to do so, make sure the PLL in FPGA is configured as Source Synchronous Mode to compensate for delay discrepancy between the clock and data (Ref [51]).

- If the FPGA core logic is running on a different clock other than the bus clock, the data has to cross the clock domain. Clock domain crossing is discussed in later chapters.

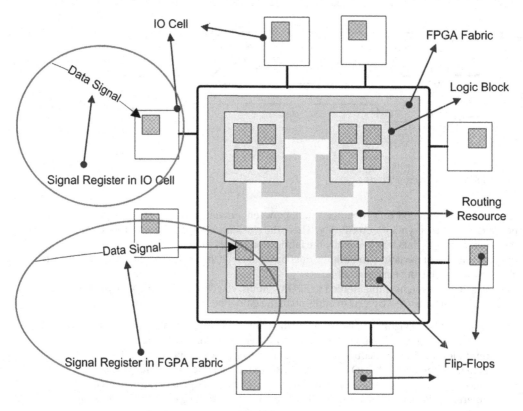

Figure 5-3. *Register signal in IO cell or FPGA fabric*

From Parallel Bus to Serial Bus

In the past 10 years, serial differential bus has gradually replaced single-ended parallel bus across the board. In the storage industry, SATA has replaced the 80-pin ribbon cable, while parallel SCSI is unseated by SAS. On the RF front, DigRF has emerged as the leading industry standard for RF transceiver. Now even the long-standing parallel PCI is succeeded by PCI-express. So this begs the question: Why is everybody jumping onto the serial bandwagon? And why, against our best intuition, does reducing the bus bit width make things better?

---

[3]Sometimes this is not a viable option, because for certain bus standards, the bus clock can be gated off when it is inactive.

The Drawbacks of Parallel Bus

At the inception of data bus evolution, bus speed was slow while PCB and IC fabrication technology were coarse, parallel bus running single-ended signals seems a reasonable design choice. And higher throughput can always be obtained by increasing the bus bit width, just like we solve the traffic jam by adding more lanes to the freeway.

However, when the bus clock frequency goes beyond 100MHz, the law of diminishing returns starts to rule. And such a approach is soon plagued by the following limiting factors:

1. *Crosstalk*

    We were told in Physics 101 that current has to flow in a closed loop, which means every signal has to find a return path for itself, and that return path is usually through the ground.[4] In the case of parallel single-ended bus, as the 3-bit wide bus shown in Figure 5-4, the three signal paths flow back in the common ground, which forms three current loops. If the common ground is a single wire, like the one used in ribbon cable, the high-frequency current loop will be inductively coupled with neighboring loops,[5] which gives rise to crosstalk.

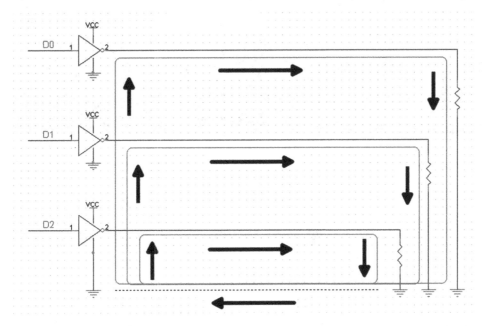

Figure 5-4. Coupled current loop

---

[4]For high-frequency signals, both power plane and ground plane can be used as the return path. But for the sake of simplicity, we assume ground plane is the closet reference plane for the rest of this book.
[5]In addition to inductive couple, there is also capacitive couple. If the return path is a uniform plane, like the ground plane in PCB, capacitive couple and inductive couple current may be in the same order of magnitude. However, if the return path is a single lead, inductive couple will dominate (Ref [20]).

To reduce the crosstalk, one thing you can do during cable fabrication is use twisted pair, which provides separate ground return path for each signal wire. In this way, each current loop will have a much smaller effective loop area. Thus, inductive coupling can be greatly reduced.

Such an approach has indeed increased the longevity of parallel ATA bus. Initially the IDE disks use 40-wire ribbon cable. However, after the introduction of Ultra ATA/33 mode (UDMA Mode 2), 80-wire ribbon cable was adopted to ensure signal integrity. The added 40-wires are all ground wires to reduce the effective loop area. And the new 80-wire cable can support a transfer rate up to 100MBps (Ultra ATA/100 mode). A photo of those two types of cables is shown in Figure 5-5. But with the non-stop drive for higher clock frequency, the added switching noise has eventually pushed parallel bus to its limit.

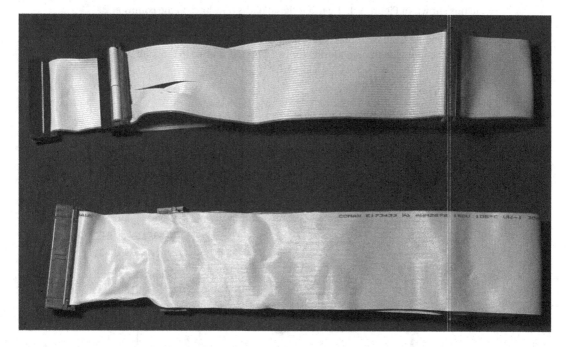

Figure 5-5. *40-wire ATA cable (top) and 80-wire ATA cable (bottom)*

2. *Skew*

People like to compare the parallel bus to a multilane highway. That leaves the illusion that throughput can always be increased by widening the bus. However, there is one subtle difference here: On a multilane highway, the vehicles in one lane do not have to go side by side with those in the neighboring lane. In other words, the flow of vehicles is asynchronous among lanes. However, on the parallel bus, the signals should stay synchronous under the command of the bus clock. Due to the layout constraint and fabrication tolerance, it is hard to keep the length of each signal trace exactly the same. And the deviation among signal traces will be aggregated when bus width increases.

The deviation among signal traces will inevitably introduce skew when signals travel on the parallel bus, as illustrated in Figure 5-6. The wider the bus is, the worse the skew will be, which puts a cap on the bus clock rate. This is another reason why higher speed busses take the form of multilane serial links instead of a wider parallel bus, as PCI express can aptly attest.

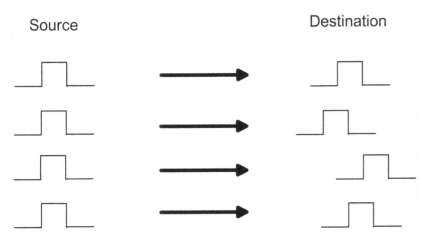

Figure 5-6. *Signal skew*

3. *Power consumption and hot swap*

   Power is also a design issue that arises when the single-ended parallel bus goes wider. The added signal traces/wires mean extra current. And the single-ended signal usually has a typical power swing like 1.8V. If terminated by a 50Ω resistor, the current will be around 36mA for each signal wire. High current is not only an issue for battery powered devices; it also poses design challenges for hot swap applications, such as disk arrays used in data centers.

The Benefit of Differential Serial Link

These technical hurdles eventually broke the camel's back on single-ended parallel bus. To look for better solutions, designers turned to the differential serial link for help.

Different from single-ended signal, the differential link sends a signal over a pair of traces/wires, as shown in Figure 5-7. Signals traveling on these two traces have the same magnitude but opposite polarity, which means the two current loops formed by them are identical except one flows in the opposite direction of the other. If you put these two traces very close to each other (as close as you can without violating layout rules), you can safely bet that their return paths will be the same as well. Since the two current loops flow in opposite directions, they will cancel each other out in the return path. So you can say that for differential pair, there is no return current flow in the ground. That is to say, the signals are all confined between the pair of differential traces/wires.

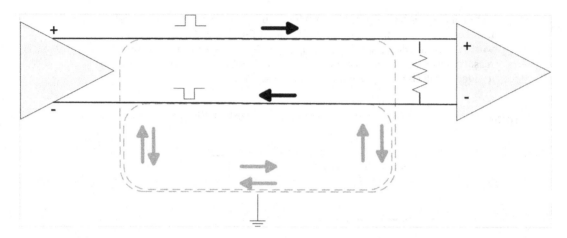

Figure 5-7. *Differential pair*

For practical purposes, the differential signal pair will also be superimposed to a common DC voltage (common mode voltage, Vcm) to keep them from clipping. And the receiver will subtract the signal pair from each other to produce the final result. In this way, the actual level of the common voltage does not much matter, as long as there is no signal clipping. An example of a differential signal level is shown in Figure 5-8.[6]

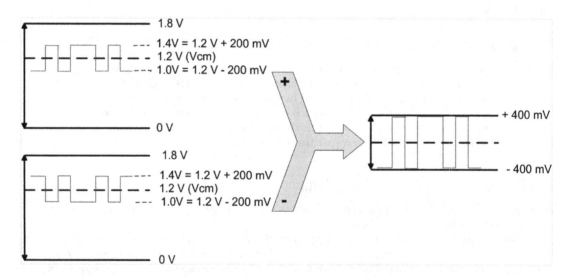

Figure 5-8. *Differential signal example*

[6]There exist a few industry standards for differential signal technology. And each has its own output swing that ranges typically from 350mV to 800mV (Ref [16]).

Now let's revisit the drawbacks of single-ended parallel bus and see how serial bus fairs in those regards:

- Since the signals are confined within the differential pair, the effective loop area is pretty small as compared to single-ended signal. In addition, due to the proximity of these two current loops, any ambient noise or nearby coupling will affect them equally as common mode noise, which will then be subtracted out at the receiver end. This means the differential pair is not susceptible to coupling from nearby traces and is less of an aggressor as far as crosstalk is concerned.

- To reduce skew, differential serial link only needs to match the length of two traces in a differential pair, and the two traces are in close proximity. So it is less subject to process variation and temperature fluctuation.

- Differential signals run at a much lower signal level. (As you can see in Figure 5-8, differential signals only have a voltage swing of a few hundred millivolts.) Due to its single bit width, differential link needs less signal traces/wires as opposed to parallel bus. And all these factors are boons for low-power design and hot swap application.

In a nutshell, you can overcome those problems by adopting differential serial link. Since the differential serial link has better signal quality, its single bit width does not stop it from running at much higher data rate.

Implementation of Differential Serial Link

Now you know that you can improve the data rate by adopting differential serial link. Since the serial link is a point-to-point connection, source synchronous clocking is an obvious choice. In that regard, there are basically two ways to send the clock along with the serial data: out-of-band clocking and embedded clocking.

Out-of-band Clocking

When the serial link data rate is in the order of a few hundred Mbps, the clock can be sent on a separate trace in single-ended mode. Both transmit pair and receive pair can use this clock as the base clock. Since the data rate is high, this single-ended clock frequency cannot be in the same order as the data rate, otherwise it will defeat the purpose of using a differential serial link. Instead, the single-ended clock is used as a base clock. Both the transmitter and receiver (for duplex communication) should use PLL to generate the corresponding internal clock frequency from the base clock, as shown in Figure 5-9.

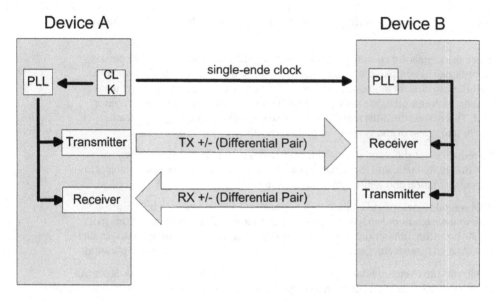

Figure 5-9. *Two differential pairs with a single-ended clock*

A typical example for such a setting is the DigRF V3 protocol, which is used to transmit/receive data samples between baseband IC and RF IC in telecomm equipment. The single-ended base clock can be 19.2MHz, 26MHz, or 38.4MHz, while the data rate on each differential pair can be as high as 312Mbps (Ref [7]).

Since the single-ended clock is well below 100MHz, the transmitter and receiver can be implemented by a FPGA that has differential IO capability. (Of course, ASIC implementation with a dedicated PHY is always possible under any circumstance.) Should you choose to do this in FPGA instead of ASIC, here is one scheme you can consider for a physical layer implementation:

1. The physical layer should define some basic frame structures for serial data. The frame can contain segments like frame type, frame length, payload, etc. At the beginning of each frame, there should be a SYNC pattern to indicate the start of the frame, as illustrated in Figure 5-10.

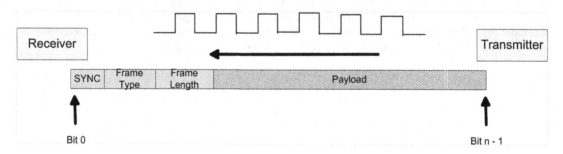

Figure 5-10. *Frame structure example*

2.   As indicated in Figure 5-9, the transmitter should derive the internal data clock by using PLL to multiply the base clock. For example, in DigRF V3 protocol, if the base clock is 26MHz, and bus data rate is 312Mbps (Ref [7]), the transmitter can use a X12 PLL to derive the 312MHz clock internally and use this clock (assume you use 0 degree phase of the PLL) to drive the P/S module (parallel to serial module, which can be implemented by a shift register).

3.   At the receiver side, you should also use PLL to generate the internal data clock from the base clock. However, due to different initial phases on both sides, the best sampling phase is unknown to the receiver. Consequently, the receiver's PLL should generate multiple phases and use a SYNC pattern detector to pick the best sampling phase, as illustrated in Figure 5-11. An algorithm should be devised to determine the best sampling phase when SYNC detectors are positive. (It is highly likely that more than one SYNC detector will report positive, so the algorithm should pick the best one among those positive detections.) Since it takes some time to get the final outcome from the algorithm, each SYNC detector has to be followed by a Frame Parser to analyze and save those frame segments that succeed the SYNC pattern. When the best sampling phase is determined, the result from the corresponding frame parser will be used for further processing.

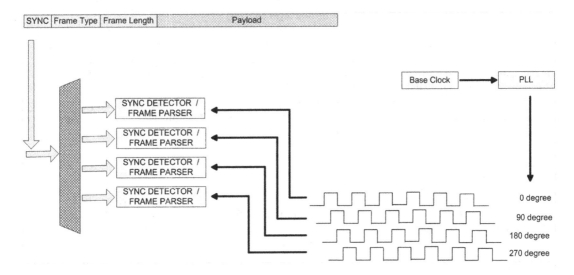

Figure 5-11. *Sampling with multiple clocks (different phase)*

Embedded Clocking

If you buy a desktop PC today, chances are that you will have a serial ATA disk inside your chassis. In the past five years, serial ATA disks have largely replaced their parallel counterparts on the desktop market. Interestingly, if you open the chassis and take a close look at your SATA signal cable, you will notice that it only has seven pins. According to the SATA spec, three of them are ground. The other four are TX+, TX-, RX+, and RX- (Ref [8]). It makes you wonder: Where is the clock?

When the serial data rate goes into the territory of Gbps, skew and jitters prevent out-of-band clocking from being a viable option. Instead, the clock is embedded into the data stream and recovered by the receiver. Detail of such scheme can be found in Refs [1] and [8]. *Since the receiver is locked to the incoming embedded clock and not to an external reference clock, jitter requirements for both transmitter and receiver input clocks are relaxed significantly* (Ref [1]).

Since the data rate is in the order of Gbps, analog behavior matters much more than digital behavior. In other words, it is pretty much out of the hands of digital designer, so a specially crafted SerDes (Serial/De-serial) module has to be used to transmit and receive data reliably. If FPGA is chosen over ASIC to implement the serial link, the FPGA device has to have a SerDes module embedded. The good news is that you don't have to go through those ordeals mentioned earlier anymore, but the bad news is now the FPGA device will cost more because of the SerDes module.

8b/10b Encoding and Signal Conditioning

In addition to the clocking scheme, 8b/10b encoding and signal conditioning are also adopted in a differential serial link to further boost the data rate.

8b/10b encoding is used to achieve DC balancing, which means that for a long stream of data bits, there are an equal number of ones and zeros in the stream. In other words, the DC component will be eliminated in the spectrum and AC coupled connection can be facilitated. 8b/10b encoding was first introduced in Ref [9], and was adopted by Serial ATA standard in its link layer (Ref [8]).

When a signal travels over the trace or cable, loss will inevitably happen, and such loss is usually not the same across the whole spectrum. Most likely, the transfer function H(x) for the trace/cable can be viewed as some sort of low-pass filter. From signal integrity's perspective, the loss of high frequency will increase the rising time for each pulse, which leads to ISI (Inter Symbol Interference) when the pulse is short. Thus it becomes critical at a high data rate to correctly model the H(x) so that it can be canceled out. Signal-conditioning techniques, such as pre-emphasis, de-emphasis and equalization, are widely adopted to do this job. The details of these techniques are beyond the scope of this book. Inquisitive readers can further explore those topics through Ref [1] and Ref [8].

Bus Standards

There's a myriad of bus standards used in the embedded systems, and this book is not trying to be an encyclopedia on that subject. However, those standards that are frequently encountered in the daily battle will be discussed in this section.

RS-232

Yes, the good old RS-232 port! The RS-232 standard has a history of more than 40 years. It is an asynchronous, single-ended bus standard. Despite all of its shortcomings, RS-232 has survived waves of tech revolution and is still actively adopted as a way to communicate between host PC and target board.

RS-232's longevity comes from its simplicity. Although its full spec has defined a connector interface of 25 pins, for all practical purposes, only three pins are needed to establish a two-way communication. On an IBM PC, RS-232 usually takes the form of a D-sub 9-pin connector (often called COM port in PC slang). Among the nine pins, you can use pin 2 (RXD), pin 3 (TXD), and pin 5 (GND) to link the target to the PC. On the target board, you only need a 3-pin header as the connector.

To make two RS-232 ports talk to each other, you might also need a null modem cable. The null modem cable makes a twist in between so that the TXD pin on one side goes to the RXD pin on the other side, and vice versa.

Often times, the term UART is used to refer to an asynchronous serial port that supports RS-232 standards. Many microcontrollers boast their abundance of peripherals by offering UART ports. However, be advised that although they might be logically compatible, the actual voltage level is different. Most UARTs that come with the microprocessors are 1.8V or 3.3V, while the actual RS-232 port could have a voltage swing between -15V and +15V. Thus a transceiver, like the one in Ref [10], is needed when you connect UARTs to an RS-232 port. The transceiver will provide voltage level translation as well as ESD protection.

When the RS-232 port (COM port) is used on an IBM PC, you need software to read/write the port. One way is to use a terminal emulator software, such as Tera Term (Ref [11]) or Hyper Terminal, to make the port a console. If you like to write your own software to control the port in Windows, you have two popular choices:

- You can use the MSCOMM control provided by Microsoft.

- You can treat the COM port as a device file and read/write it directly. This requires more coding as well as skills for multi-thread programming, but it gives you the best flexibility.

I2C and SMBus

I2C bus was first introduced by Philips Semiconductors (now NXP Semiconductors) as a two-wire bus standard for device interconnection. It uses two signal wires operating in OD (Open Drain) mode to achieve bi-directional communication between the master and slave devices. Thanks to its succinctness, I2C bus was widely used as serial interface for memory, temperature sensor, enclosure LED controller, etc. And later Intel made a derivative standard out of it called SMBus (Ref [13]), which is used on PC motherboard. There are some minor deviations between those two standards, but for all practical purposes, they can be treated the same way.

The two wires used in I2C and SMBus are named SCL (Serial Clock Line) and SDA (Serial Data Line). Since they both operate in OD mode, they are set to high by default through pull-up resistors. Each slave device should be programmed with a 7-bit address to uniquely identify itself on the bus (a 10-bit address is also possible although not widely used (Ref [12])), and the whole data transaction includes stages like START, ADDRESS, R/W, ACK, DATA, STOP, etc. that can be found in Ref [12] and Ref [13]. Most of the time, developers can skip those protocol details if the microprocessor has an I2C controller built-in or the FPGA contains IP cores for I2C.

However, if your product is sensitive to cost, or if you just can't find enough I2C controllers, you might consider using GPIOs to toggle the SDA and SCL. Since I2C bus often works on a 100KHz clock[7] while modern CPUs run with a clock rate at least hundreds of MHz, you should have enough CPU horsepower to keep up with the I2C bus. Thus CPU can use a timer to control and drive the GPIOs. The drawback of this approach is that you now have to get into the protocol detail and implement everything, from the START/STOP condition to the slave address, in the software. But as you can imagine, the software does not count into the BOM cost. :-). Because of this, such an approach is generally called *bit banging* in embedded lingo, and it is not confined to I2C implementation only.

And to help the development of I2C bus, a host bus tool like the one in Ref [49] can be used as a good reference.

---

[7]There are other clock frequencies defined by the standard, like 400KHz, or even 1MHz and 3.4MHz. Make sure you have done the number crunching before you take this approach.

SPI

SPI (Serial Peripheral Interface) is a 4-wire, full-duplex serial bus protocol introduced by Motorola's semiconductor division (Now part of NXP). The four wires are named SCLK (Serial Clock), SS# (Slave Select, Active Low), MOSI (Master Output, Slave Input), and MISO (Master Input, Slave Output). The SS# serves as a chip select to indicate the availability of data frame, and two-way traffic can flow on MISO and MOSI at the same time when SS# is asserted. The timing diagram between an SPI master and an SPI slave device is illustrated in Figure 5-12.

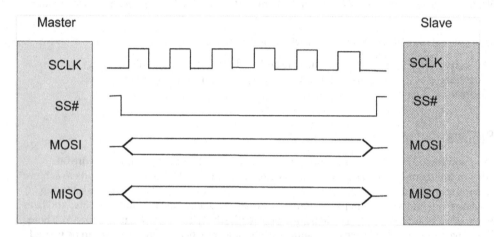

Figure 5-12. *SPI bus*

The SPI bus offers the flexibility to change clock polarity (CPOL) and clock phase (CPHA), which determines the actual sampling edge. Details about these definitions can be found in Ref [14]. However, unlike other bus standards, SPI does not have a central committee to craft out all the subtle details. So the actual implementation from different vendors may have slight discrepancies on the frame length or the SS#.[8] FPGA or a small CPLD might be necessary to resolve such discrepancies.

And to help the development of SPI bus, a host bus tool like the one in Ref [49] can be used as a good reference.

PCI overview

PCI and many of its variants (CompactPCI, PCI-104, PCIe-104, etc.) are the main expansion bus on the IBM PC and embedded computers. On an IBM PC, PCI started out in parallel form as a PCI Local Bus (Ref [4]) to replace its predecessor (ISA and EISA bus). PCI Local Bus runs at a clock of 33MHz or 66MHz, with the bit width of 32-bit or 64-bit. PCI Local Bus uses level triggered interrupt, and the latest PCI Local Bus also specifies MSI as a design option (Ref [4]). With the emergence of the high-speed serial bus, PCI local bus is gradually being replaced by its serial successor—PCI Express (Ref [3]). On an IBM PC, PCI bus is connected to South Bridge. However, the South Bridge itself might also have ports for USB, Ethernet, IDE/SATA, etc. So some devices, like the IDE/SATA disks, could be connected to South Bridge directly. Or they could be hooked up to a HBA adapter on the PCI bus, as illustrated in Figure 5-13.

---

[8]Some vendors define SS# as a pulse instead of a level and define a fixed frame length.

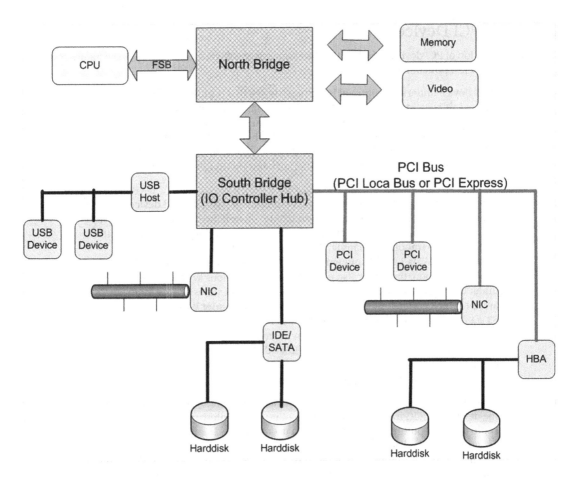

Figure 5-13. *PCI bus on IBM PC*

PCI Local Bus supports auto configuration, which is essential to plug-and-play application. Any PCI device that requires memory spaces or IO spaces should implement base address registers in its configuration space. During power up and system initialization, system software (BIOS for IBM PC) will enumerate each PCI bus segment, inquire each PCI device by writing all 1s to the base address register and read it back, and then fill the base address register with the assigned starting addresses (Ref [4]). The device driver should operate within the assigned address spaces. Thus no address conflict would happen at the system level.

As mentioned, PCI Local Bus is gradually being replaced by its serial successor PCI Express, which has a multilane, point-to-point architecture. PCI Local Bus and PCI Express are very complicated bus standards, and they run a wide gamut of topics, ranging from electrical spec to mechanical spec, and many of the physical layer details have to be dealt by the ASIC or IP core. So this book will leave their hardware design part to inquisitive readers. Refer to Refs [4] and [3] or the corresponding ASIC manual for more detail. As for software development, configuration space is what matters most, since that is what device driver is based on. And PCI Express is backward compatible in this regard (Ref [3]).

Driver for PCI Device

The implementation of PCI devices involves the development of software. Depending on the nature of the device, there are a few possibilities:

- If the device under development is a critical device that is indispensable during system initialization, such as a HBA (Host Bus Adapter) for disk storage, the bootloader has to find a way to identify and communicate with the device before OS is loaded. In the case of an IBM PC, it means that some BIOS code has to be developed for this device besides the OS driver so that system BIOS could talk to this new device for OS loading. BIOS is a different beast on IBM PC and this book will leave this part to inquisitive readers.

- The OS also needs drivers to identify and communicate with the PCI device. Again, if the device is indispensable during system boot up or if the device is IO intensive by its nature, a full-blown driver should be developed to reduce the interrupt latency and guarantee smooth OS loading. For Windows developers, it usually means working with WDK or DDK and writing a driver for kernel mode.[9] In addition to the normal operation, the driver also needs to deal with various system power states under the ACPI standard (Ref [16]). The most important states among them are S3 and S4, which correspond to standby and hibernation under Windows.[10] For the S3 state, the content of the main memory will be kept intact while for the S4 state everything is flushed to the disk and main memory will lose its power as well. This could be a challenge for HBA disk driver since disk storage will be the first device to come up before everything else can be restored from disk. The driver has to make sure it reinitializes everything correctly when it is woken up from hibernation.

- However, if the device is not a critical device needed during system initialization, and the device is not IO intensive (i.e, it can tolerate longer interrupt latency), developers could consider putting most of the code in user mode and leaving only a small, generic part in the kernel mode, as illustrated Figure 5-14. The plus side of such an approach is that developers can enjoy more latitude in the user mode (such as the abundance of debugging tools) without worrying about corrupting the system core. In fact, there are tools out there, such as WinDriver (Ref [15]), to assist you in taking this rapid development approach (Microsoft WDK also provides user mode driver support through UMDF—User Mode Driver Framework). When the driver code is matured, it can be migrated into kernel mode for better performance.

---

[9]If the device already has a corresponding driver type under Windows, such as Network Adapter or SCSI, developers only need to write a miniport driver to handle the hardware specific operation. Windows will handle the rest through its bus driver.
[10]Windows allows you to choose standby or hibernation when you shut down the computer. Sometimes the hibernation option may not be available because there are missing drivers in the system. (You will see a yellow exclamation icon in the Device Manager.) An incorrect setting in system BIOS may also affect these options.

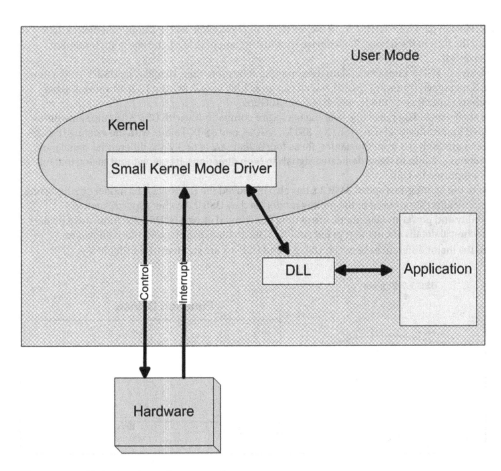

Figure 5-14. *Drive hardware in user mode*

USB overview

Anyone who owns a PC would probably agree that USB (Universal Serial Bus) is the most convenient way to connect peripheral devices to it. Since its debut in 1996, USB has continuously sharpened its specs and won great popularity. At the time when this book is being conceived, USB 3.1 (Ref [46]) is the most up-to-date USB spec, although many devices still stick with USB 2.0. So this book covers both the USB 3.x and USB 2.0 specs here. However, for all practical purposes, this book will focus on USB function devices only and leave USB host and USB hub devices to inquisitive readers.

USB 2.0 is a differential serial bus with half-duplex mode only. Its interface has four signal wires: VBUS, D+, D-, and GND (Ref [17]). USB 2.0 supports three data rates: high speed (480Mbps), full speed (12Mbps), and low speed (1.5Mbps). The device can be bus-powered or self-powered. If the device is bus-powered, such as the popular USB flash drive, the maximum current it can draw from the bus is 500mA (USB 3.x supports up to 900mA for a bus-powered device).

Unlike other bus standards, such as PCI Local Bus, USB host is the only bus master on the USB 2.0 bus. In other words, there is no active notification from USB device to USB host under the USB 2.0 standard. Every transaction is initiated by a USB host. On the device side, the device should provide buffers for data read and write. Such a buffer is called the *endpoint* in USB lingo. Since a USB bus is host centered, an IN endpoint refers to data transferring from the device to the host while an OUT endpoint means data transferring from the host to the device. Each device should implement at least one endpoint (endpoint zero) for device control.

On top of USB 2.0, USB 3.x has cranked up the throughput by more than 10 fold. The USB 3.0 has a new data rate called SuperSpeed (not warp speed), which can go as high as 5Gbps. And USB 3.1 raises the bar even higher, with its SuperSpeed+ mode, which reaches 10Gbps.

To achieve such a high data rate while keeping backward compatibility with USB 2.0 at the same time, USB 3.x has added four extra signal wires (SSTX+, SSTX-, SSRX+, and SSRX-) along with the existing USB 2.0 signals. As you can probably tell from the names, these four signals are actually two differential signal pairs in opposite directions. Thanks to those dedicated signals in both directions, it goes without saying that full duplex mode is supported by USB 3.x.

In addition to the lighting-fast speed, USB 3.x has also improved the USB data flow model by supporting asynchronous notification as opposed to the polling model used by USB 2.0. Other aspects, such as streaming capability and power management, are getting fined tuned as well by USB 3.x. Inquisitive readers can find more technical details about them in Ref [18][46]. The architecture of a USB bus is shown in Figure 5-15, and the major differences between USB 2.0 and USB 3.x are summarized in Table 5-1.

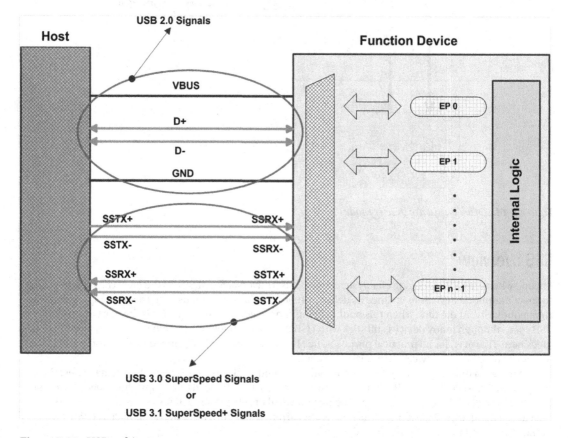

Figure 5-15. USB architecture

Table 5-1. *Major Differences Between USB 2.0 and USB 3.x*

	USB 2.0	USB 3.x
Highest Data Rate	*High speed (480Mbps)*	*USB 3.0 Super Speed (5Gbps)* *USB 3.1 Super Speed+ (10Gbps)*
Full Duplex Mode	*Not supported*	*Supported*
Asynchronous Notification	*Not supported*	*Supported*
Maximum Current for Bus-powered Device	*500 mA*	*900 mA*
Data Signals	*D±*	*D±, SSTX± and SSRX±*

A regular USB device can only communicate with a USB host. In other words, regular USB devices are unable to talk among themselves without the coordination of a USB host. However, there are application scenarios where USB devices do need to connect to each other directly, such as a digital camera downloading photos to a portable printer. USB OTG (Ref [21]) is thus devised as a supplement to USB 2.0 to cover those special cases. A USB OTG device can act as a host temporarily and recognize only a limited number of devices on its list. For more information, refer to Ref [21]. This book will not discuss USB OTG in any further detail.

Independent of the data interfaces (USB 2.0/3.x), standards are also defined for power delivery through USB. The USB power delivery specification (Ref [47]) attempts to address the increasing demands for supplying power directly through USB. The maximum voltage and current defined in Ref [47] is 20V and 5A each, which can deliver a power up to 100 watts. (With that much of juice, the USB cable is probably the only wire your printer needs.) And with USB power delivery, it also means the power can flow both ways (i.e., the laptop can be both a power supplier and a power receiver under various circumstances).

In addition to the type A/B connector defined by USB 2.0/3.x, a new connector type (Type-C) has also been introduced along with USB 3.1, which has a symmetric mechanical design and automatic side/pin detection (Ref [48]). The Type-C cable can deliver maximum current of 3A, which can be used for both data and power delivery. And it seems Type-C has the ambition to unseat all the other cable/wires for your laptop.[11]

USB device implementation

The implementation of the USB device covers a wide gamut of topics, including hardware components, USB firmware (all sorts of USB descriptors), Windows device drivers, etc. All of them are indispensable in the making of a successful USB device. This book will touch all of them to present a whole picture to the reader.

---

[11]Apple made a bold decision to include only one single Type-C port in its new MacBook, which debuted on April 10th, 2015.

Selecting the Hardware Component

Basically, this means you need to choose an IC solution for the USB interface. The available choices are:

1. Using a separate USB MAC and USB PHY (transceiver)

   If the CPU vendor has already integrated USB Link controller inside the processor, then a discrete USB PHY (transceiver) is the only thing you need. (Most likely, it will be a ULPI interface (Ref [22]).) Actually most CPU vendors would also provide or recommend companion chips to match their CPUs, which probably contain the USB transceiver. One of such examples is TI OMAP 35x processor. It has a USB link controller integrated inside. At the same time, its companion PMIC has a USB transceiver integrated (Ref [24]). And it is always better to follow the application notes and pick the recommend solution from the CPU vendor if there is one.

2. Using an off-the-shelf USB/UART solution

   If all you need is a quick and straightforward USB communication between the host and the device, you might consider using off-the-shelf USB/UART solution, such as the one mentioned in Ref [25]. The generic architecture of such solution is shown in Figure 5-16. On the device side, it is pretty much the same as other USB devices: internal logic will communicate with the off-the-shelf IC through FIFOs (or parallel bus) to read/write endpoints. However, the USB descriptor and the host device driver are designed so that the USB device appears as a virtual RS-232 COM port on the host side. As mentioned, the good old RS-232 port has been used for such a long time that there are tons of mature applications you can choose from, and it is not difficult to make your own COM port application. However, the simplicity that comes with RS-232 is the biggest downside of this approach.

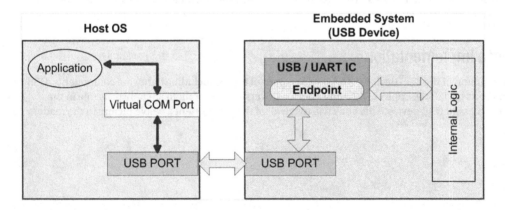

Figure 5-16. *USB/UART*

3. Using a USB microcontroller

    USB microcontroller, such as the one mentioned in Ref [26], is good for the situation where a full-blown USB device implementation is needed. It usually includes three parts in one chip: USB transceiver, USB link controller, and a generic microprocessor core, as illustrated in Figure 5-17. The USB link controller will handle all the low-level protocols, such as packet decoding, while the microprocessor core will deal with the high-level protocols, like providing all kinds of USB descriptors and handling USB requests. Consequently, a lot of firmware work has to be done for the microprocessor core. In addition, software is also needed at the host side. The device driver has to be developed for the host OS. On the Windows platform, WinUSB can be used to replace the device driver if there is only one application talking to the USB device. The technical details of the USB firmware, device driver, and WinUSB will be discussed in later sections.

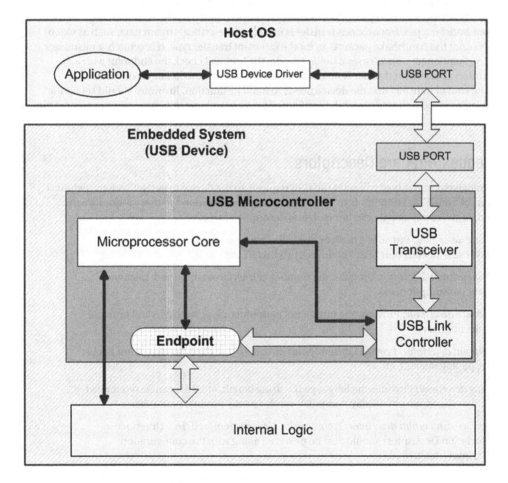

Figure 5-17. *USB microcontroller*

Implementing the USB firmware

The USB hardware usually needs to be configured by the firmware. Other than the control flow, the firmware is expected to finish the following jobs

Setting Up Endpoints

Each USB transfer is carried out by exchanging a series of USB packets between the host and device. A single USB packet can be the type of token, data, handshake, etc. (Ref [17][18]). However, those low-level packets are usually taken care of by hardware, and they are not firmware's concern most of the time. Instead, firmware works at the granularity of USB transfer. There are four main types of USB transfer defined by USB standard: control transfer, bulk transfer, interrupt transfer, and isochronous transfer. Accordingly, there are four types of endpoints to match those transfer types. Endpoint also has its direction, whereby IN refers to the transfer from device to host and OUT means the transfer from host to device.

Each transfer type has its own sequence of packets that are predefined by the USB standard. Note that among those four transfer types, isochronous transfer is used for time-critical stream data, such as video or audio, and does not use handshake packets. As for the interrupt transfer type, it is actually a misnomer under USB 2.0. As mentioned, USB 2.0 uses a polling mode; the host will check the endpoint with a predetermined interval. Interrupt transfer is meant to be used for small, low-latency data transfers.

Based on the kind of transfers that the device needs to fulfill its function, firmware should set up the endpoints. And endpoint zero should always be configured for a control transfer type, as designated by the USB standard.

Handling Request/Prepare Descriptors

During device initialization, the host will send the USB requests to the device to get/set various technical parameters (control transfer to endpoint zero), and the device should respond to these requests with corresponding data structures called *descriptors*. These descriptors include:

- *Device descriptor:* Provides a device class/subclass, a protocol, and a maximum packet size for endpoint zero, vendor ID, product ID, etc.

- *Configuration descriptor:* Provides the number of interfaces supported, maximum power consumption, etc.

- *Interface descriptor:* Provides the number of endpoints, class/subclass, and protocol for the interface.

- *Endpoint descriptor:* Provides the endpoint direction, transfer type, maximum packet size, polling interval, etc.

- *String descriptor:* Provides the language ID, string length, etc. This can be referenced by other descriptors to display things like vendor name, product name, etc.

- *Interface association descriptor:* If composite device is required, IAD (Interface Association Descriptor) should also be provided along with the configuration descriptor (Ref [27][28]).

In addition, for a specific device class, other descriptors, such as function descriptors, are also needed. The detailed format of these descriptors can be found in Ref [17] and [18]). In the companion material of this book, there is a complete example of CDC device (USB communication device), which contains a generic head file that has all the data structure for standard USB descriptors and all the class code definitions. This head file can be useful for other USB firmware projects.

Implementing the USB Device Driver (for the Host OS)

USB devices do not work by themselves. Instead they are designed to communicate with USB hosts. If you happen to own a PC that runs Windows, you probably know that the "Found New Hardware Wizard" will appear when you plug in your USB device the first time. In other words, the driver has to be provided at that point in order to let the USB device become a functional peripheral under the host OS. Due to the dominance of Windows on the PC desktop, this book will focus this discussion on the Windows platform only. As far as the USB device driver is concerned, there are several options:

- *Writing a full-fledged device driver:* You can always be hardcore and write a full-fledged Windows driver for the device. However, I can tell you first hand that it is not a small undertaking to write a Windows driver. For those strong hearts, refer to WDK and MSDN for more information. Of course, if the situation does warrant a full-fledged driver, you have to bite the bullet. In that case, I would suggest you start with an existing example in WDK and go from there.

    But under some circumstances, you might find more lightweight solutions, such as USB Virtual COM port or WinUSB.

- *Using a USB Virtual COM port:* One thing you can do to circumvent the development of a full-fledged Windows driver is use virtual COM port. The USB device can present itself as an analog modem type of PSTN device. And PSTN device (Ref [30]) falls under a more generic USB class called CDC device (communication device, Ref [29]). The best part for such a setting is that Windows already has a built-in device driver called usbser.sys for these kinds of modem devices. All you have to do is to write an .inf file, which gives the corresponding vendor ID and product ID and instructs Windows to load usbser.sys for the device. All the heavy lifting will be done by usbser.sys behind the scenes.

    Such a USB CDC device will have IN and OUT bulk endpoints for transfer in both directions. And all the legacy applications that previously communicate with COM port can still be used seamlessly. As mentioned, off-the-shelf IC solutions can also be found among various vendors (such as those mentioned in Ref [25]). So it is truly a cost-effective solution as long as the COM port is acceptable in the actual circumstances. The companion material in this book has a full examples of the USB Virtual COM port implementation based on the Cypress CY3684 FX2LP EZ-USB Development Kit.

- *Using WinUSB:* Oftentimes the manufacturer of the USB device is also the software provider of the Windows application. That application usually is the only one that talks to the USB device. If that is the case, and if there is no isochronous endpoint on the USB device, WinUSB might be a good choice rather than developing a full-fledged Windows driver (Ref [19]).

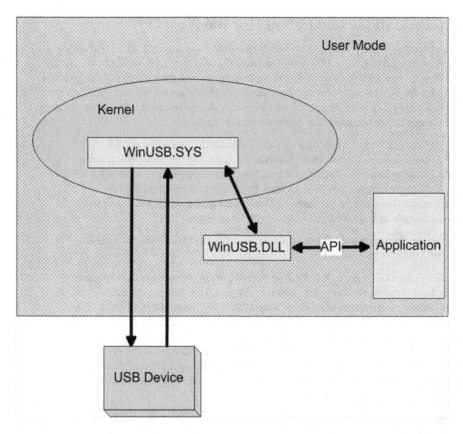

Figure 5-18. *WinUSB architecture*

As illustrated in Figure 5-18, the WinUSB solution is made up of two main components: WinUSB.SYS in kernel mode and WinUSB.DLL in user mode. WinUSB.SYS is a generic driver that communicates with the USB device. WinUSB.DLL is the user mode component that talks with WinUSB.SYS and provides APIs to the user applications. All the device details are thus masked out by those APIs, which makes the user application more portable and easier to develop. Come to think of it, all USB devices are nothing but a bunch of endpoints after all, so such generic solution is entirely feasible, as long as throughput and latency are acceptable under practical circumstances.

The companion material of this book has a complete example of a WinUSB implementation based on the same USB Virtual COM port device.

Ethernet

Ethernet started in the early days of Internet and it has been evolving ever since. This discussion on bus standards wouldn't been complete without it.

Overview

The explosive growth of the Internet in the past two decades has given Ethernet a universal presence. The data rate of Ethernet has also skyrocketed from the initial 10Mbps to the order of 100Gbps. Thus it becomes more and more imperative for embedded systems to be able to support Ethernet connections. In fact, the plummeting hardware prices and the soaring CPU speed have even made it practically possible for medium-scale systems to run the entire web server inside the microprocessor so that configuration can be done easily through the browser, as demonstrated in Figure 5-19.

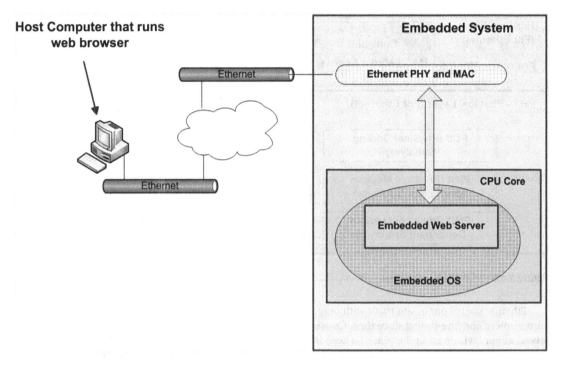

Figure 5-19. *Embedded web server*

The network protocols that shore up the whole architecture are implied in Figure 5-19. "Network protocol" usually refers to the TCP/IP protocol stack that forms the bedrock of the whole Internet. As shown in Figure 5-20, TCP/IP comprises multiple layers of protocols and Ethernet resides in the lower two layers: the Physical Layer and the Data Link Layer. Each layer is also composed of several sublayers. As far as Ethernet hardware is concerned, the sublayers marked by hatched lines in Figure 5-20 are usually implemented in silicon while the rest of the protocol stack is done by the software.

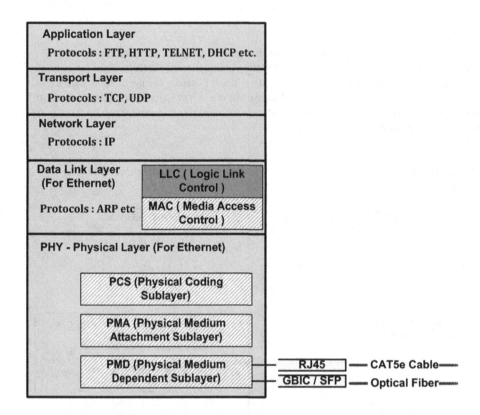

Figure 5-20. *TCP/IP over Ethernet*

Ethernet started out in late 1970s with only 10Mbps. It has gone through multiple generations of improvement and fine-tuning since then. Consequently, the physical layer of Ethernet has also gone through several major revisions over the years. So keep in mind that the structure shown Figure 5-20 is based mainly on Fast Ethernet (100Mbps) and Gigabit Ethernet (1000Mbps) for the sake of practicality. Different sublayer structures may be found for other Ethernet standards, such as the 10Mbps Classic Ethernet.

Among the three subphysical layers shown in Figure 5-20, the PCS layer is mainly responsible for encoding the data coming from MAC into a form that is suitable for physical transmission. Since most Ethernet link are AC-coupled, the encoded bit stream should be DC-balanced. Also, for the sake of clock sync, the encoded bit stream should also have a short run length, which means the maximum number of consecutive 1 and 0 after encoding should be reasonably small. In 10Mbps Classic Ethernet, Manchester code is used in PCS and is replaced by the 4B/5B code scheme in Fast Ethernet (100Mbps), and then is succeeded by the 8B/10B code scheme (Ref [9]) in Gigabit Ethernet (1000Mbps).

PCS will send the encoded bit streams to the next layer, called PMA, through a standard interface. In Gigabit Ethernet, TBI (Ten Bit Interface) or RTBI (Reduced Ten Bit Interface) are usually the ones adopted. PMA is mainly responsible for data serialization and de-serialization.

Bit streams processed by PMA will be passed along to PMD, which is deeply tied to the actual physical medium being used in data transferring. There are a wide variety of medium being used or once used in the Ethernet, such as coaxial cable with BNC connector, twisted pair with RJ45 connector, optical fiber with GBIC/SFP module. For all practical purposes, the rest of this book will focus on the discussion of xBase-T Ethernet protocol, which uses RJ45 connector and UTP (Unshielded Twisted Pair) cable. This is the most popular cable/connector you'll often find on your desktop PC or laptop computer.

The interface between MAC and PHY has also been sharply enhanced and standardized over the years to accommodate for surging data rates. The possible interfaces are AUI or SNI for 10Mbps Classic Ethernet; MII or RMII for 100Mbps Fast Ethernet; GMII, RGMII, or SGMII for 1000Mbps Gigabit Ethernet; XGMII for 10Gigabit Ethernet; etc. This topic will be revisited in later sections that discuss the hardware configuration.

Hardware Implementation

As mentioned, MAC and PHY (PCS + PMA + PMD) are usually implemented in IC while the rest of the TCP/IP is realized by the software. As for the IC implementation, there are two solutions, discussed next.

Standalone Ethernet Controller

One design choice is to have both the MAC and PHY in one package and make a standalone controller out of it. The controller will have a generic bus interface or processor interface for control/configuration and data transferring. The popular choices are:

- *Using a 10Mbps standalone controller:* If all you need is a cheap Ethernet link that could get you onto the network, and if data rate is not a major concern, you might use a mature 10Base-T solution, such as the one in Ref [31]. Since 10Mbps Ethernet is from the dinosaur age, most its controllers carry ISA 16-bit interface and can be tweaked easily to communicate with microprocessors. The plus side is that since these are mature solutions, there are application notes and proven designs galore. The flip side is that 10Mbps is slowly dying out. My two cents is to use it only as an interface for debug and development and never populate it to the final product.

- *Using a 10/100Mbps standalone controller:* For medium-scale embedded systems, 100Mbps Fast Ethernet link offers a nice balance between cost and performance. A standalone 10/100Mbps controller, such as those in Ref [32][33][34], can be used if the microprocessor does not have a built-in MAC. Most 10/100Mbps controllers will have ISA bus interfaces or generic parallel bus interfaces for integration with microprocessor or FPGA, and due to the popularity of these controllers, main-stream embedded OS usually have built-in drivers ready to recognize them as valid peripherals.

- *Using a standalone controller for 1000Mbps Gigabit Ethernet:* There are many vendors offering 1000Mbps standalone controllers on the market. However, due to the high throughput of Gigabit Ethernet, most of these controllers have a PCI/PCI-express bus interface. And it might post a design challenge if the PCI/PCI-express is not natively supported by the microprocessor used in the system. An alternative is to use a microprocessor that has built-in Gigabit MAC or FPGAs with a Gigabit MAC core, and a Gigabit PHY as an external transceiver. This brings you to the second solution: separate MAC and PHY.

Separate MAC and PHY

Another possible design choice is to use a separate Ethernet MAC and PHY. For IC implementation, the PHY is usually designed as a standalone transceiver due to its analog nature. The MAC could be a standalone IC by itself, or it could be a built-in module that comes with the microprocessor. It is also possible for FPGAs to carry a MAC core if necessary.

Since now the MAC chip and PHY chip could be provided by two different vendors, before any Ethernet data can be exchanged between MAC and PHY, a standard interface has to be defined for MAC chip to configure the PHY registers and manage the PHY chip. The industry standard interface for this purpose is called MDIO, which is part of the MII and GMII interface. MDIO is a serial bus interface, with two signals— MDC (clock, no more than 2.5MHz) and MDIO (bi-directional serial data)—as illustrated in Figure 5-21.

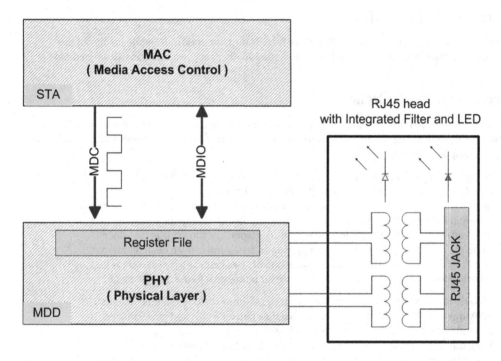

Figure 5-21. *MDIO interface*

There is only one PHY in Figure 5-21. However, the MDIO interface is actually defined with the management of multiple PHYs in mind. And in MDIO lingo, the entity sending out the clock is called STA (Station Management Entity) and the one being managed is called MDD (MDIO Manageable Device), which corresponds to MAC and PHY respectively in Figure 5-21. Each PHY should be assigned an address during board design to identify itself to the MAC. This address will be contained in the MDIO frame structure when MAC configures the PHY. The details of MDIO frame structure and the definition of MII management register can be found in Ref [40].

Despite the functional partition shown in Figure 5-22, PCS and PMA sometimes can be moved into a MAC module at implementation due to their digital nature, and PHY chip could focus on the analog transceiver to simplify the design and manufacture. PCS and PMA can also be supported by both the MAC chip and PHY chip. Thus, the possible partition of work (hardware configuration) could be as following after initialization:

1. PCS and PMA stay with PHY chip

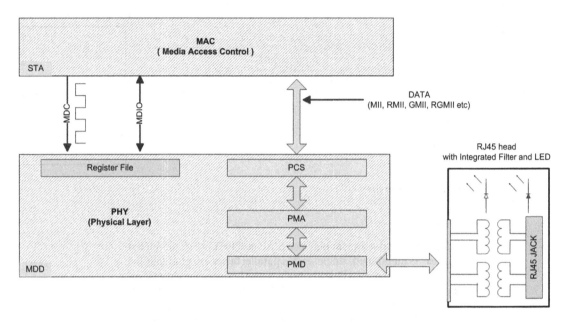

Figure 5-22. *PCS and PMA implemented by the PHY module*

As illustrated in Figure 5-22, MAC and PHY can exchange data through MII/RMII interface for Fast Ethernet, or GMII/RGMII interface for Gigabit Ethernet. These interfaces are parallel bus interfaces, with separate TX and RX clocks. The RX clock is always provided by PHY module while the TX clock could be provided either way depending on the Ethernet Protocol. (The TX clock is from PHY under Fast Ethernet. But under Gigabit Ethernet, MAC is responsible for providing the TX clock.).Details of the signal pins can be found in Refs [37][40].

2. PCS stays with MAC and PMA stays with PHY

It is also possible to let the MAC chip implement the PCS layer while PMA stays with PHY, as illustrated in Figure 5-23. If such a configuration is chosen under Gigabit Ethernet, data can be exchanged between PCS and PMA through a TBI or RTBI interface. Both TBI and RTBI are parallel bus interfaces, with PCS providing the TX clock and PMA providing the RX clock. Details of the signal pins for TBI/RTBI can be found in Refs [37][40].

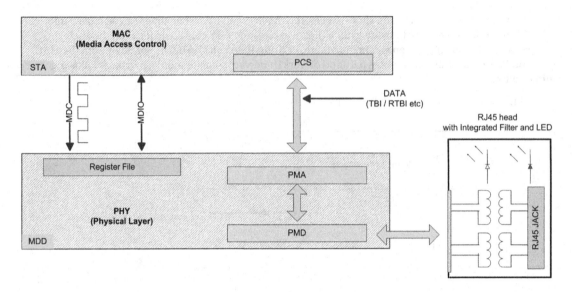

Figure 5-23. *PCS in MAC; PMA in PHY*

3. SGMII interface

    All the interfaces discussed so far between MAC and PHY are single-ended parallel buses. As mentioned, single-ended parallel buses are gradually being unseated by serial differential links throughout the industry. Ethernet is also following this trend with no exceptions. A serial protocol called SGMII (Ref [41], including 10/100/1000Mbps Data Rate) has been introduced by Cisco and won the support of major IC vendors. The idea behind SGMII is to use SERDES to serialize TBI bus shown in Figure 5-23 into serial links, recover the GMII data at the other end, and follow the path of Figure 5-22 afterward. So conceptually there will be PCS modules in both MAC and PHY, as illustrated in Figure 5-24.

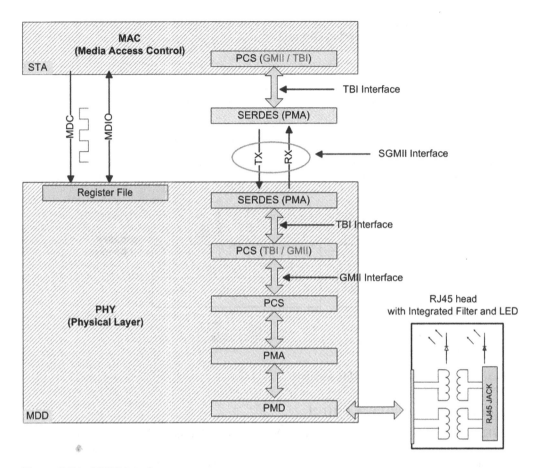

Figure 5-24. *SGMII interface*

The SERDES that converts the parallel TBI bus into a serial link is shown as a component external to MAC in Figure 5-24. Depending on the actual silicon implementation, sometimes the SERDES can also be absorbed into a MAC module to further simplify board design. As for the clock, there is no out-of-band clock for the serial link shown in Figure 5-24. That is to say, the clock is embedded in the serial data link. However, out-of-band clock implementation is also possible under SGMII protocol if that is desired by designers (Ref [41]).

4.   PCS and PMA stay with MAC; PHY has PMD only

If 1000Mbps Gigabit Ethernet is the only protocol that needs to be supported, Figure 5-24 can be simplified with PMD left in the PHY alone. In this case, PHY turns into a standalone Ethernet transceiver. For an Ethernet switch, such an implementation has its merits since a Gigabit Ethernet switch port might be connected to copper wire (RJ45, twisted pair) or optical fiber of various wave lengths. To allow maximum flexibility, the PMD and the corresponding connector can be mechanically made into a standard form factor module, like GBIC or SFP. Due to such a standard form factor, a mixture of fiber and copper port connection is possible under the same switch core fabric, as illustrated in Figure 5-25.

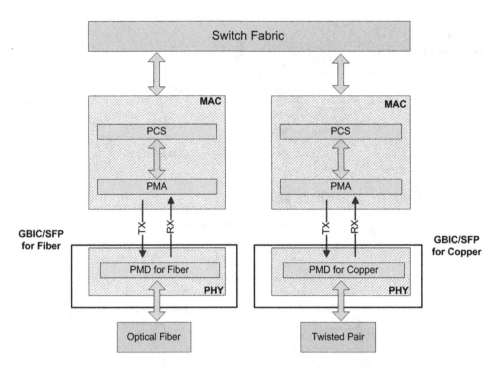

Figure 5-25. *GBIC/SFP with a common MAC interface*

Ethernet Software and Flow Control

To work with Ethernet at the software level, most embedded systems will adopt the TCP/IP protocols provided by the OS. The protocol stack will interact with MAC through Ethernet Frames. In that sense, at the software level there is no major difference between an embedded system and a desktop PC.

However, sometimes there might be a throughput limitation imposed by physical layer, which is common among many embedded systems. Under such circumstances, flow control problems will arise when the board is connected to a powerful PC or switch that blows data relentlessly. Pause frame is one of the ways to throttle the traffic.

1. Enable flow control for the PC or switch. For Windows PC, right-click the adapter in Device Manager and choose Property. Look for things like "flow control" or "performance option". For switch, it is often configured through web interface.

2. On the embedded system side, enable pause frame in both the MAC and PHY. Set a big pause quanta, such as 0xFFFF, to begin with.

3. In MAC, set the FIFO threshold that triggers the pause frame. The initial threshold can be set at the middle point of FIFO size.

4. Use wireshark (Ref [50]) to observe and capture the pause frame. The filter can be set as `mcc.opcode == 0x1`.

Summary

The bus is what stitches everything together.

There are main two kinds of bus timing: system synchronous timing and source synchronous timing. As the bus clock rate keeps going up, source synchronous timing starts to gain in popularity. These days, high-speed buses are moving away from parallel single-ended links into differential serial links. Depending on the clock rate, there are multiple ways to implement the interface for a differential serial bus.

Major bus standards—including RS-232, I2C, SPI, PCI, USB, and Ethernet—were also deeply explored in this chapter.

References

1. LVDS Owner's Manual (4th Edition). National Semiconductor, 2008
2. "Best Practices for the Quartus II TimeQuest Timing Analyzer." (Chapter 7, *Quartus II Handbook*. Version 10.1, Vol 3)
3. PCI Express Base Specification 3.0, PCI-SIG, November 10, 2010
4. PCI Local Bus Specification Revision 3.0, PCI-SIG, August 12, 2002
5. *Signal Integrity Simplified*. Eric Bogatin, Prentice Hall Professional Technical Reference, 2004
6. *Signal Integrity Issues and Printed Circuit Board Design*. Douglas Brooks, Prentice Hall Professional Technical Reference, 2003
7. Datasheet of Agilent N4850A DigRF v3 Acquisition Probe, N4860A DigRF v3 Stimulus Probe, Agilent Technologies, Inc., 2007
8. Serial ATA Revision 3.0, Serial ATA International Organization, May, 2009
9. A DC-Balanced, Partitioned-Block, 8B/10B Transmission Code, A. X. Widmer and P. A. Franaszek, *IBM Journal of Research and Development*, Vol. 27, No.5, September, 1983
10. ±15kV ESD-Protected, 1μA, 3.0V to 5.5V, 250kbps, RS-232 Transceivers with AutoShutdown (Rev 6), Maxim Integrated Products, Inc., September, 2005
11. Tera Term Open Source Project (http://ttssh2.sourceforge.jp)
12. UM10204, I2C-Bus Specification and User Manual Rev. 03, *NXP Semiconductors*, June 19, 2007
13. System Management Bus (SMBus) Specification Version 2.0, SBS Implementers Forum, August 3, 2000
14. AN-300, SPI Bus Compatibility, FM25160 16KB SPI FRAM, Ramtron International Corporation, February, 1999
15. *WinDriver PCI/ISA/CardBus User's Manual,* Version 10.2.1, Jungo Ltd., 2010
16. Advanced Configuration and Power Interface Specification, Rev 4.0a, Hewlett-Packard Corp, Intel Corp, Microsoft Corp, Phoenix Tech Ltd, Toshiba Corp, April, 5, 2010
17. Universal Serial Bus Specification, Revision 2.0, Compaq, Hewlett-Packard, Intel, Lucent, Microsoft, NEC, Philips, April 27, 2000
18. Universal Serial Bus 3.0 Specification, Revision 1.0, Hewlett-Packard Company, Intel Corporation, Microsoft Corporation, NEC Corporation, ST-NXP Wireless, Texas Instruments, November 12, 2008
19. "How to Use WinUSB to Communicate with a USB Device," Microsoft Corporation, March 9, 2010
20. EZ-USB Technical Reference Manual, Version 1.2, Cypress Semiconductor Corporation, 2005
21. On-The-Go and Embedded Host Supplement to the USB Revision 2.0 Specification, Revision 2.0 plus errata and ecn, USB Implementers Forum, Inc., June 4, 2010
22. UTMI+ Low Pin Interface (ULPI) Specification, Revision 1.1, ULPI Working Group, October, 20, 2004
23. UTMI+ Specification, Revision 1.0, February 25, 2004
24. TPS65950 Integrated Power Management/Audio Codec, Texas Instruments, SWCS034A, May 2008
25. FT232R USB UART IC, Data Sheet Version 2.13, Future Technology Devices International Ltd, 2015
26. CY7C68013A/CY7C68014A, CY7C68015A/CY7C68016A EZ-USB FX2LP™ USB Microcontroller, Cypress Semiconductor Corporation, 04/18/2005

27. USB ECN, Interface Association Descriptors
28. USB Interface Association Descriptor Device Class Code and Use Model, Intel Corporation, Rev 1.0, July 23, 2003
29. Universal Serial Bus Class Definitions for Communications Devices, Rev 1.2, November 3, 2010
30. Universal Serial Bus Communications Class Subclass Specification for PSTN Devices, Rev 1.2, February 9, 2007
31. Crystal LAN™ CS8900A Ethernet Controller Technical Reference Manual, AN83REV3, Cirrus Logic, Inc. June 2001
32. DM9000 ISA to Ethernet MAC Controller with Integrated 10/100 PHY, Version DM9000-DS-F03, DAVICOM Semiconductor, Inc., April 23, 2009
33. DM9000 32/16/8 bit Three-In-One Fast Ethernet Controller Application Notes V1.22, Version DM9000-AP-1.22, June 11, 2004
34. LAN9221/LAN9221i, High-Performance 16-bit Non-PCI 10/100 Ethernet Controller with Variable Voltage I/O, Rev 2.8, Standard Microsystems Corp. (SMSC), July 14, 2011
35. DP83848C PHYTER®—Commercial Temperature Single Port 10/100 Mb/s Ethernet Physical Layer Transceiver, National Semiconductor Corporation, May 2008
36. DP83848T PHYTER® Mini-Industrial Temperature Single 10/100 Ethernet Transceiver, National Semiconductor Corporation, May 2008
37. Triple-Speed Ethernet MegaCore Function User Guide, Software Version: 11.0, Altera Corporation, June 2011
38. 88E1111 Product Brief, Integrated 10/100/1000 Ultra Gigabit Ethernet Transceiver, Marvell Semiconductor, Inc., March 4, 2009
39. Use The MDIO Bus To Interrogate Complex Devices, Electronic Design, December 3, 2001
40. IEEE Std 802.3™-2008, IEEE Standard for Information technology—Telecommunications and Information Exchange Between Systems—Local and Metropolitan Area Networks—Specific requirements Part 3: Carrier sense multiple access with Collision Detection (CSMA/CD) Access Method and Physical Layer Specifications, IEEE Computer Society, December 26, 2008
41. Serial-GMII Specification, Document Number ENG-46158, Revision 1.8, Yi-Chin Chu, Cisco Systems, November 2, 2005
42. Intel 8080 Microcomputer Systems User's Manual, September 1975
43. MC6800 8-bit Microprocessing Unit (MPU), Motorola Semiconductor Products Inc., 1984
44. M6800 Programming Reference Manual, First Edition, Motorola, Inc., 1976
45. AN10880, Using the LCD Interface in the LPC313x/LPC314x/LPC315x Devices, Rev.01, NXP Semiconductors, October 21, 2009
46. Universal Serial Bus 3.1 Specification, Rev 1.0, Hewlett-Packard Company, Intel Corporation, Microsoft Corporation, Renesas Corporation, ST-Ericsson, Texas Instruments, July 26, 2013
47. Universal Serial Bus Power Delivery Specification, Revision 2.0, V1.1. 7, May, 2015
48. Universal Serial Bus Type-C Cable and Connector Specification, Revision 1.1, April 3, 2015
49. Aardvark I2C/SPI Host Adapters User Manual V5.15, Total Phase, Inc., February 28, 2014
50. Wireshark (https://www.wireshark.org/)
51. Phase Locked Loop (ALTPLL) Megafunction User Guide, Altera Corporation, November, 2009
52. *Digital Integrated Circuits: A Design Perspective (2nd Edition)*. Jan M. Rabaey, Anantha Chandrakasan, Borivoje Nikolic, Pearson Education, Inc., 2003
53. *Silicon VLSI Technology: Fundamentals, Practice and Modeling*. James D. Plummer, Michael D. Deal, Peter B. Griffin, Prentice Hall, Inc., 2000

CHAPTER 6

Firmware Coding in C

When I read commentary about suggestions for where C should go, I often think back and give thanks that it wasn't developed under the advice of a worldwide crowd.

—Dennis Ritchie (1941 – 2011), the Father of the C Language

Starting in this chapter, I will delve into the softer side of embedded systems. The next three chapters cover C/C++, the build process, and a little bit of embedded OS.

Overview

At the time when this book was being conceived, iPhone was on its way to become one of the coolest gadgets in history. And a large part of its success comes from the contribution of app developers worldwide. (Of course, there is no doubt that iPhone also owns its success to Steve Jobs' charisma and vision. And yes, we are all gonna miss him.) Due to the far-reaching deployment of handheld devices like the iPhone and the eye-catching nature of cool apps (as that angry bird can attest), people start to intuitively take those apps as the equivalence of embedded software, which is *not true* technically.

As illustrated in Figure 6-1, applications are not the only software in embedded systems. Instead, multiple layers of software exist above the bare bones hardware but below the applications. These layers of low-level software, such as the device driver and the embedded OS, form the foundation for applications. In the world of embedded systems, people often use the term *firmware* to distinguish them from the high-level application software. Firmware runs at a level that is very close to the hardware, and it has to deal with a lot of primitive constraints that are not felt by high-level applications.

© Changyi Gu 2016
C. Gu, *Building Embedded Systems*, DOI 10.1007/978-1-4842-1919-5_6

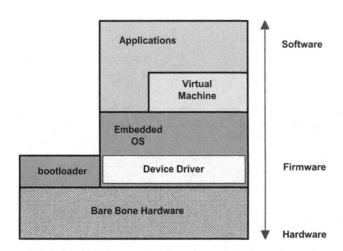

Figure 6-1. software, firmware, and hardware

The boundary between firmware and pure software is not clean-cut, thanks to the relative nature of terms like "low" and "close". However, my personal observation is that things like bootloader, device driver, and embedded OS are traditionally recognized by the industry as firmware instead of software. After all, these pieces have to be customized for each individual hardware platform while the application software is usually cross-platform. In fact, such a difference is also reflected in the development flow, as most application software can be developed on a resource-rich Windows/PC platform (or Mac or Linux host) and ported to the embedded system later without much effort. But for firmware developers, the flow usually involves reading hardware specs and working with ICE or debug console in the lab, so the development environment is much cruder for those folks.

If I have to use analogy, imagine the whole embedded system as a race car. The car itself, of course, is the hardware that includes various components like engine, wheels, etc. The application developers are more like the driver of the race car. Their job is to rev up the engine, spin the wheels, drive the car to full throttle, and roar pass other cars on the track. If they are lucky enough to win, they get to be wooed by fans and pop open the champagne bottle. Yes, that is as cool as cool can be. However, firmware folks are more like the mechanics who take care of the car. They have to worry about the oil leaks somewhere in the valves. They have to replace the worn-out tire before it blows. They need to make sure the engine never overheats at full speed. In a nutshell, their job is to fine-tune the machine to its maximum. Oftentimes their hands are dotted with oil patches. They don't have much time under the limelight.

Now, at this point, if you are doubting your career choice, it is still not too late to throw in the towel... :-). Aw, who am I kidding? In fact, I was a firmware engineer in my previous life. I can tell you first-hand that firmware is cool in its own way. It is the bridge between software and hardware, and you get to have innumerable tricks up your sleeve when you work for both sides!

The Confusion of Tongues

After the Tower of Babel collapsed, varieties of languages mushroomed across the earth. People who don't speak the same language started to have a difficult time understanding each other, and confusion and frustration followed. In the world of embedded systems, firmware engineers are challenged by the same dilemma. What is the best programming language for firmware development? As of today, the possible choices are:

- Assembly

- C

- C++

- Java

Due to the plethora of programming languages, this list is by no means to be complete. (For example, someone from the Microsoft camp might point out that C# is a good alternative to Java.) However, as mentioned, firmware has to have close interaction with the hardware. In that sense, any language that is based on a virtual machine is not suitable for firmware development, since it naturally begs the question: "What programming language shall we choose to write the virtual machine?" So this basically leaves both Java and C# out. (Of course, they can still be valuable assets to application developers.)

Among all the choices, assembly language is native to the hardware and it can be employed to drive the hardware to its maximum efficiency. However, its biggest strength is also its Achilles heel. Since it is closely tied to the hardware platform, portability is the major concern for assembly language. Assembly is hard to master due to its primitive nature. It lacks the abstract data structures that are commonly enjoyed by modern high-level programming languages (like C or C++). All these deficiencies make Assembly justifiable only in some special cases like:

- The startup code for bootloader.

- Part of the C runtime library to set up the stack and do other initializations.

- In places where execution time is extremely critical. One such case is the FFT (Fast Fourier Transform) for digital signal processing. It is usually worth the effort to optimize the FFT with assembly language under such circumstances, since FFT is often involved in the core algorithms of telecomm, image processing, audio/video compression/decompression, etc. And those applications generally demand realtime processing.

- In places where code size has to be reduced to meet the space constraints. (In these days, Flash becomes the commodity and compilers are doing a good job on optimization. So before you start the size-reduction with Assembly, make sure all the other options have been seriously explored.)

So the choice finally comes down to C or C++. There has been plenty of heated debates and discussions on this subject in the past decade or so. Here are a few interesting observations I have made personally during my career:

- OS kernel developers tend to have a penchant for C. Mr. Linus Torvalds is probably the most prominent member among this group of folks. He once made a few strong remarks against using C++ in Linux Kernel (Ref [2]). On the other hand, engineers at Microsoft seem to have their reservations as well when it comes to writing a kernel mode driver in C++. By reading Ref [1] and [2], I found it hard to believe that, for the first time in as long as I can remember, Microsoft and the Linux community were able to agree on something at the fundamental level.

- Although my personal opinions could be biased (as the nature of all subjective matters), I want to make it clear that I hold no prejudice for one language over the other. I think OO is an ingenious concept, and it is possible for C programmers to utilize that idea without actually using C++, as demonstrated in Refs [3][4].

- Despite the dominance of C in the kernel, I have to admit that ISOCPP[1] has done a great job refining C++ in the past decade. And embedded system developers can definitely benefit from embracing C++ under viable circumstances.

The rest of this chapter discusses firmware coding in the C language, followed by another chapter on C++.

Firmware Coding in the C Language

As mentioned, for any embedded systems, Assembly will play a small part after reset to initialize the hardware, set up the stack, zero out the .bss section, and jump to the main() function. After that point, everything is carried out in the C language, except those special cases where Assembly coding is still needed for the sake of performance/sizing.

It is also worth mentioning again that firmware has its own idiosyncrasies, most of which are not a concern for application development. But these idiosyncrasies will easily break the firmware if they are not done in the right way. I will try to address them as much as I can in this section.

Explicit Type of Bit-Width

One of the main functions of the firmware is to configure hardware, which usually involves register/memory access. Most register and memory buses will have a bit-width that is the multiple of a byte (8-bit). Since each CPU has its own native word length (bit-width) for bus and register file, the size of integer-type varies from one compiler to another. It is always good practice for a firmware project to typedef its own primitive types, with explicit bit-width marking. Those "typedef" should be put in a common head file and included by the whole project to avoid any ambiguity on bit-width, as exemplified in Listing 6-1. In Listing 6-1, 8-bit, 16-bit, and 32-bit unsigned integer types are defined for the KEIL C51 compiler and the GNU C 32-bit compiler. (Notice that GNU C can support 64-bit operations on 32-bit hardware through a software library. So 64-bit unsigned integer can be defined as "long long" type in GNU C.) All the rest of the project should only use these U8/U16/U32/U64 types to define and declare variables for the sake of readability and portability. DSP programmers might also want to define the signed integer type (S8/S16/S32) for their own platforms if necessary.

Note that there are some C_ASSERT macros used in Listing 6-1. C_ASSERT is a debug macro defined for compile-time assertion, and its purpose is to statically catch anything fishy. It is discussed in detail later in this chapter.

Listing 6-1. Explicit Type of Bit-Width

```
#ifndef COMMON_TYPE_H
#define COMMON_TYPE_H

#include "debug.h"

#if defined (KEIL_C51)
```

---

[1]ISOCPP stands for C++ Standards Committee, the international standardization working group for C++.

```
    typedef unsigned long int    U32;
    typedef unsigned short       U16;
    typedef unsigned char        U8;

...

#elif defined (GNU_C_32BIT)

    typedef long long       U64;
    typedef unsigned int    U32;
    typedef unsigned short  U16;
    typedef unsigned char   U8;

    C_ASSERT(sizeof(U64) == 8);

...

#else

...

#end if

C_ASSERT(sizeof(U32) == 4);
C_ASSERT(sizeof(U16) == 2);
C_ASSERT(sizeof(U8) == 1);

#endif
```

Align the Data Structure

As mentioned in previous section, each CPU has its own word length (bit-width) and it varies from one platform to another. Due to the broad scope of embedded CPUs, the word length can be as small as 8-bit (such as Intel 8051), or as big as 32-bit (like the popular ARM processor). For some high-end products, they could even afford 64-bit CPUs. And most systems would also allow memory access with a width less than the native word length (For example, most 32-bit CPUs can R/W memory in a single byte as well, with degraded efficiency.) Consequently, it gives rise to the alignment issue for all data structures.

From the performance standpoint, memory accesses that are aligned to native word length are always preferred (i.e., the beginning address is at the boundary of native word length, and the R/W length is a multiple of native word length.) Thus most compilers will do some optimization by inserting space padding into the data structure when they see the chance of misalignment, as demonstrated in Listing 6-2.

Listing 6-2. Padding for Alignment

```
typedef struct {
    U8 a;
    U32 b;
    U16 c;
} STRUCT_NOT_PACKED;
```

GNU C Compiler, 32 bit target CPU:

sizeof(STRUCT_NOT_PACKED) == 12

For a 32-bit target CPU, the GNU C compiler will insert three bytes of padding for field a and two bytes padding for field c in the struct defined in Listing 6-2, so that struct member b will be aligned to 32-bit boundary, and the total size of struct STRUCT_NOT_PACKED becomes 12.

Under GNC C, such alignment padding can be disabled by using the packed attribute, as illustrated in Listing 6-3. After applying the packed attribute, the size of the data structure is reduced to 7. However, struct member b is now stored across 32-bit boundary and carries performance penalties if accessed individually.

Listing 6-3. Packed Data Structure

```
#define PACKED __attribute__((packed))

typedef struct {
    U8  a;
    U32 b;
    U16 c;
} PACKED STRUCT_PACKED;
```

GNU C Compiler, 32 bit target CPU:

sizeof(STRUCT_PACKED) == 7

Alignment is an important part of firmware programming, and most hardware controllers will assume the data structures they need are aligned in memory with a predefined layout. For example, DMA controllers will read a scatter-gather list out of the memory when it starts DMA, and for the sake of efficiency, it usually requires such a list being stored in memory with boundary alignment and having a certain format. Thus it is the firmware engineer's job to lay out the data structure correctly per the hardware specs.

To deal with data alignment issue and improve portability, firmware engineers can borrow a page from Microsoft's playbook. In fact, I found the following two macros from Ref [5] handy in this regard:

- FIELD_OFFSET: This macro is defined in Listing 6-4. It can be used to indicate the actual offset of a member field within the struct. With the help of this macro, you can get that the member field b is offset by 4 bytes and 1 byte each in the previous two structs defined by Listings 6-2 and 6-3.

Listing 6-4. FIELD_OFFSET Macro

```
#define FIELD_OFFSET(type, field) \
        ((long)(long*)&(((type *)0)->field))
```

GNU C Compiler, 32 bit target CPU:

```
FIELD_OFFSET(STRUCT_NOT_PACKED, b) == 4
FIELD_OFFSET(STRUCT_PACKED, b) == 1
```

- TYPE_ALIGNMENT: As shown in Listing 6-5, TYPE_ALIGNMENT[2] can be used to determine the alignment offset of a data structure. And if used together with C_ASSERT, it can set up the first line of defense to sort out alignment issues at compile time.

Listing 6-5. TYPE_ALIGNMENT Macro

```
#define TYPE_ALIGNMENT( t ) \
    ((long)(sizeof(struct { char x; t test; }) - sizeof(t)))
```

GNU C Compiler, 32 bit target CPU:

```
C_ASSERT(TYPE_ALIGNMENT(STRUCT_PACKED) == 1);
```

```
C_ASSERT(TYPE_ALIGNMENT(STRUCT_NOT_PACKED) == 4);
```

As a strong proponent of code craftsmanship, I also encourage firmware engineers to manually pad the data structure and reorder the member fields for better portability and less space consumption, assuming the data structure is user-defined.

For example, the struct in Listing 6-2 and Listing 6-3 can be hand-optimized into something like what's shown in Listing 6-6. With an alignment of 32-bit, the new struct is more portable across different compilers and compile options. And its size is reduced to 8 instead of 12.

Listing 6-6. Manual Padding for Alignment

```
typedef struct {
    U8  a;
    U8  padding;
    U16 c;
    U32 b;
} STRUCT_MANUAL_PADDING;
```

GNU C Compiler, 32 bit target CPU:

```
sizeof(STRUCT_MANUAL_PADDING) == 8
```

```
TYPE_ALIGNMENT(STRUCT_MANUAL_PADDING) == 4
```

Debug Print

With the rapid betterment of JTAG-based ICE, firmware engineers now get to trace their code at source level and enjoy all the benefits that are brought by the IDE (Integrated Development Environment, such as Eclipse or ARM RVDS) tools. The productivity of firmware development can thus leapfrog with the assistance of these tools. My opinion is that generally they should take the top of the list during project budgeting.

---

[2]TYPE_ALIGNMENT is defined differently in Ref [5] as FIELD_OFFSET(struct { char x; t test; }, test). The side effect of this implementation is that compiler will give out warnings if it is used together with C_ASSERT.

However, despite the prowess of these ICE and IDE tools, Debug Console is still an indispensable alternative during firmware development. Sometimes, it might be the only viable option:

- To achieve source level debugging, the object has to be built with all the symbols in. Oftentimes it also requires disabling the compiler optimization by using -O0 option. All these treatments will bloat the object size. Under some circumstances it might be infeasible to load the object through ICE after these treatments, due to the constraint of memory size.

- Thanks to the different level of optimization used for compiling debug build and release build, there could be inconsistency on code flow and latency. Consequently, certain bugs may appear in one build but not the other. For release build, Debug Console probably is the only feasible option to save engineers from total fiasco.

- For device drivers, some of them can be dynamically loaded instead of statically built-in. For example, Linux kernel modules can be loaded dynamically by using the insmod command after the kernel is fully booted up. In order to trace these modules through IDE tools, developers have to jump through a lot of hoops. And this is usually the moment when printk() comes handy.

For hardware implementation, Debug Console is traditionally done by using UART ports available on the microprocessor, with external RS-232 transceivers for voltage-level translation and ESD protection. Alternatively, Debug Console can also be implemented through a JTAG port.

As for firmware, it means that you have to define a macro to print out debug information on the console. This macro can be disabled in the release build, and it should support variable number of arguments, just like the printf() or printk() function. Here are the common recipes that I have collected over the years for such debug print macros:

- *Brutal force for a variable number of arguments:* Basically, this means using multiple macros for different number of arguments, as shown in Listing 6-7. This is not an elegant way of doing things, but the code is very straightforward and portable. The drawback is, of course, that you have to use a different macro every time you need to print out more arguments.

Listing 6-7. Multiple Macros for Debug Print

```
#if defined(DEBUG)
    #define DEBUG_OUTPUT(val) \
            printk(val)

    #define DEBUG_OUTPUT2(val1,val2) \
            printk(val1,(int)(val2))

    #define DEBUG_OUTPUT3(val1,val2,val3) \
            printk(val1,(int)(val2),(int)(val3))
#else
    #define DEBUG_OUTPUT(val)
    #define DEBUG_OUTPUT2(val1,val2)
    #define DEBUG_OUTPUT3(val1,val2,val3)
#endif
```

- *Borrow a page from script playbook:* Script languages, such as Perl and Tcl, often use a debug flag to control the debug output. This trick can be borrowed by C as well, as shown in Listing 6-8. Although this brings the benefit of runtime flexibility, it may also bring runtime overhead and other side effects, depending on the compiler optimization.

Listing 6-8. Using a Debug Flag

```
extern U8 debug_flag;

#define DEBUG_PRINT if(debug_flag)printf
```

Or

```
#define DEBUG_FLAG (0)

#define DEBUG_PRINT if(DEBUG_FLAG)printf
```

- *Using double parentheses:* Another way to support a variable number of arguments is to use double parentheses whenever Debug Print is invoked, as shown in Listing 6-9. This trick is often played by Windows driver developers, but it is also applicable to firmware development. The good news is that such practice conforms to standard C, so it should be acceptable to C compilers across the board.

Listing 6-9. Using Double Parentheses

```
#if defined(DEBUG)
    #define DEBUG_PRINT(_x_) \
            do{printf ("DEBUG: "); printf _x_;}while(0)
#else
#define DEBUG_PRINT(_x_)
#endif
```

Double Parentheses:

```
DEBUG_PRINT (("arg1 = %d, arg2 = %d \n", 1, 2));
```

- *Variable number of arguments in GNU C:* If you feel the double parentheses in Listing 6-9 are bothering you, you can use a better solution if your compiler is GNU C. GNU C has native support for macros that have a variable number of arguments, as demonstrated in Listing 6-10. The double parentheses are no longer needed in this case, but the flip side is that it is not standard C practice and may have portability issues with other compilers.

Listing 6-10. Variable Number of Arguments in GNU C

```
#if defined(DEBUG)
    #define DEBUG_PRINT(fmt, args...) \
            printf("DEBUG: " fmt, ##args)
#else
    #define DEBUG_PRINT(fmt, args...)
#endif
```

```
No need for Double Braces:

DEBUG_PRINT ("arg1 = %d, arg2 = %d \n", 1, 2);
```

Compile Time Assertion

Assertions are frequently used by firmware engineers and application developers to verify the assumptions they have taken in the code flow, which is a practice strongly encouraged by most coding guidelines. However, those assertions that stay in the code flow are runtime assertions, and they are activated only when the code is executed. In addition to those runtime assertions, it is also desirable to have a way to verify assumptions at compile time, even before the code is executed. For example, when you want a compile-time assertion to do sanity checks statically.

Fortunately, there is a way to do this in the C language, and I highly recommend it to firmware practitioners. It was first introduced by Mr. Dan Saks in Ref [6], and was later enhanced by Mr. Ken Peters and Mr. Mike Teachman. The idea is to generate negative array indexes if the assertion fails during prepossessing, and therefore raise a compile error, as illustrated in Listing 6-11. The C_ASSERT macro defined in Listing 6-11 can be adopted to verify the size of a data type, the alignment of data structures, etc., whose usage has already been demonstrated in previous sections.

Listing 6-11. Compile Time Assertion

```
#define C_ASSERT(cond) \
    extern char compile_time_assertion[(cond) ? 1 : -1]
```

```
Example:

C_ASSERT(sizeof(U32) == 4);
C_ASSERT(sizeof(U16) == 2);
C_ASSERT(sizeof(U8) == 1);
```

Volatile, Const, and Peripheral Registers

"Volatile" and "Const" are often used to decorate pointer variables and raw addresses in C. However, misuse of these type qualifiers could lead to unintended consequences.

Type-Qualifier "volatile"

As we all know, RAM access is much slower than register access. If you know beforehand that a certain variable will be used frequently, you would like it to be stored in register to speed up subsequent accesses. One way to achieve such acceleration in C is to specify the storage class of that variable as register instead of auto, so that compiler will store the variable in a register if there is one. However, such manual optimization has its limitations and is usually unnecessary in practice. Instead, you could let compiler figure out the best register-allocation scheme by turning on the optimization switch, and most modern C compilers do a good job in this regard.

■ **Labs**   Let's do some experiments to get an idea on how compilers will optimize the memory access. The results shown here are produced with a 32-bit Cygwin on a 64-bit x86 machine running Windows 10. The object file format is pei-i386. I chose 32-bit Cygwin since most embedded processors are 32 bits or fewer. If you run it with 64-bit Cygwin, the object file format will be pei-x86-64, and the final Assembly code will use 64-bit instructions. But the same idea of optimization applies.

Listing 6-12. Test Program: no_volatile.c

```
unsigned int gVar = 0xABCD1234;

int main()
{
    int i;
    unsigned int *p = &gVar;
    unsigned int k = 0;

    for (i = 0; i < 100; ++i) {
        k += (*p);
    }

    return k;
} // End of main()
```

LAB 6-1: OPTIMIZATION ON MEMORY ACCESS

As shown in Listing 6-12, pointer p is pointing to global variable gVar, and (*p) will be accumulated on k 100 times. Now let's compile it without any optimization and do a disassembly.

```
$> gcc -g -O0 no_volatile.c -o no_volatile
$>
$> objdump -S -s no_volatile
```

If it is done correctly, objdump will give you the following:

...

Disassembly of section .text:
...

```
  int i;
  unsigned int *p = &gVar;
4010ba: c7 45 f8 00 20 40 00    movl    $0x402000,-0x8(%ebp)

  unsigned int k = 0;
4010c1: c7 45 f4 00 00 00 00    movl    $0x0,-0xc(%ebp)

  for (i = 0; i < 100; ++i) {
4010c8: c7 45 fc 00 00 00 00    movl    $0x0,-0x4(%ebp)
4010cf: 83 7d fc 63             cmpl    $0x63,-0x4(%ebp)
4010d3: 7f 11                   jg      4010e6 <_main+0x56>

    k += (*p);
4010d5: 8b 45 f8                mov     -0x8(%ebp),%eax
4010d8: 8b 10                   mov     (%eax),%edx
4010da: 8d 45 f4                lea     -0xc(%ebp),%eax
4010dd: 01 10                   add     %edx,(%eax)
{
    int i;
    unsigned int *p = &gVar;
    unsigned int k = 0;

    for (i = 0; i < 100; ++i) {
4010df: 8d 45 fc                lea     -0x4(%ebp),%eax
4010e2: ff 00                   incl    (%eax)
4010e4: eb e9                   jmp     4010cf <_main+0x3f>
        k += (*p);
    }

    return k;
4010e6: 8b 45 f4                mov     -0xc(%ebp),%eax

} // End of main()
...
```

Contents of section .data:
```
 402000 3412cdab 00000000 00000000 00000000  4...............
 402010 00000000 00000000 00000000 00000000  ................
 402020 00000000 00000000 00000000 00000000  ................
 402030 00000000 00000000 00000000 00000000  ................
```

...

As you can see from the bold lines (4010d5 - 4010dd), (*p) and k are both accessed directly from memory without any optimization in the loop body. (Both p and k are stored on the stack frame, as indicated by 4010ba - 4010c1. The address of gVar is 0x402000, and it was stored in p as an immediate number in 4010ba. More details about the GNU x86 assembly can be found in Ref [7] if you are not family with the syntax.)

Now if you turn up the optimization by one notch (from -O0 to -O1) and do the previous steps again, you will get the following:

```
$> gcc -g -O1 no_volatile.c -o no_volatile
$>
$> objdump -S -s no_volatile


...


Disassembly of section .text:
  ...

  int i;
  unsigned int *p = &gVar;
  unsigned int k = 0;
4010a8: b8 00 00 00 00              mov     $0x0,%eax

  for (i = 0; i < 100; ++i) {
4010ad: ba 00 00 00 00              mov     $0x0,%edx
4010b2: 8b 0d 00 20 40 00          mov     0x402000,%ecx
      k += (*p);
4010b8: 01 c8                       add     %ecx,%eax
{
    int i;
    unsigned int *p = &gVar;
    unsigned int k = 0;

    for (i = 0; i < 100; ++i) {
4010ba: 42                          inc     %edx
4010bb: 83 fa 63                    cmp     $0x63,%edx
4010be: 7e f8                       jle     4010b8 <_main+0x28>
      k += (*p);
    }

    return k;
} // End of main()
  ...

Contents of section .data:
 402000 3412cdab 00000000 00000000 00000000  4...............
 402010 00000000 00000000 00000000 00000000  ................
 402020 00000000 00000000 00000000 00000000  ................
 402030 00000000 00000000 00000000 00000000  ................


  ...
```

The major improvement on memory access is demonstrated by the bold lines in 4010b2 - 4010b8, and is summarized as following:

- First of all, the value of gVar is moved into register ecx in 4010b2 (Notice that there is no dollar sign in front of the address 0x402000. So it is direct memory addressing. The value stored at address 0x402000 is moved into ecx.)

- Secondly, the loop body now has only one instruction (0x4010b8), which does the accumulation only on registers (adding ecx to <u>eax</u>).

Thus the compiler optimization used register access to replace memory access for both k and (*p), and it therefore saved 200 memory accesses.

As you can see from Lab 6-1, the compiler has optimized away almost all memory accesses and replaced them with register accesses. However, there is one catch: *All bets are off if the memory location that p is pointing to can be modified externally.* If that is the case, replacing memory access with register access could have harmful side effects, which in turn lays the ground for inconsistency between the debug build and the release build.

For example, in Lab 6-1, the global variable gVar might be modified by ISR (Interrupt Service Routine), or pointer p could point to some memory-mapped registers that belong to a peripheral sitting on the bus. If the possibility of external modification exists for certain memory locations, such knowledge should be conveyed to the compiler as well. Otherwise, harmful optimizations might be carried out on them inadvertently.

Fortunately, C provides the type-qualifier of volatile for this purpose, as demonstrated in Lab 6-2.

■ **Labs**   Assuming the value of gVar can be changed externally in Listing 6-12, let's see what the type-qualifier volatile can do to make things right. In Listing 6-13, the keyword volatile was added when pointer p is declared.

Listing 6-13. Test Program: volatile.c

```
unsigned int gVar = 0xABCD1234;

int main()
{
    int i;
    volatile unsigned int *p = &gVar;
    unsigned int k = 0;

    for (i = 0; i < 100; ++i) {
        k += (*p);
    }

    return k;
} // End of main()
```

LAB 6-2: TYPE-QUALIFIER "VOLATILE"

If you compile Listing 6-13 with the -O1 optimization, you get the following objdump:

```
$> gcc -g -O1 volatile.c -o volatile
$>
$> objdump -S -s volatile


...

Disassembly of section .text:
  ...
  int i;
  volatile unsigned int *p = &gVar;
4010a9: bb 00 20 40 00          mov    $0x402000,%ebx


  unsigned int k = 0;
4010ae: b9 00 00 00 00          mov    $0x0,%ecx

  for (i = 0; i < 100; ++i) {
4010b3: ba 00 00 00 00          mov    $0x0,%edx
     k += (*p);
4010b8: 8b 03                   mov    (%ebx),%eax
4010ba: 01 c1                   add    %eax,%ecx
{
   int i;
   volatile unsigned int *p = &gVar;
   unsigned int k = 0;

   for (i = 0; i < 100; ++i) {
4010bc: 42                      inc    %edx
4010bd: 83 fa 63                cmp    $0x63,%edx
4010c0: 7e f6                   jle    4010b8 <_main+0x28>
     k += (*p);
   }

   return k;
} // End of main()

  ...

Contents of section .data:
 402000 3412cdab 00000000 00000000 00000000  4...............
 402010 00000000 00000000 00000000 00000000  ................
 402020 00000000 00000000 00000000 00000000  ................
 402030 00000000 00000000 00000000 00000000  ................

  ...
```

From the bold lines above, you can come away with the following observations:

- First, access to (*p) is now conducted on memory instead of register (Notice there is a dollar sign in front of address 0x402000 in 4010a9, so it is moving an immediate number 0x402000 into ebx. And in 4010b8, register indirect addressing is used to move the value stored at 0x402000 into register eax.

- Secondly, as you can tell from 4010ae and 4010ba, variable k still gets optimized by placing it in register ecx.

The volatile keyword did its job by stopping the compiler from optimizing (*p).

Unlike the register keyword, which specifies the storage class of a variable, the volatile keyword is a type qualifier. The purpose of volatile is to stop the compiler from optimization on memory access, and it is not responsible for the allocation of variable's storage space. In fact, it is possible to put both register and volatile in the same variable declaration, as illustrated in Listing 6-14.

Listing 6-14. register_volatile.c

```
unsigned int gVar = 0xABCD1234;

int main()
{
    register volatile unsigned int *p = &gVar;
    unsigned int k = 0;

    k += (*p);
    k += (*p);

    return k;
} // End of main()
```

If you compile Listing 6-14 in GNU C without any optimization, you will see that pointer p is stored in register instead of memory, but all accesses to (*p) are still conducted directly on memory.

Note that the volatile keyword can be placed before or after the type that it is qualifying. However, if the variable is a pointer, placing volatile right next to the pointer could mean something different, as illustrated by the following examples:

- Non-volatile pointer that points to volatile data:

  ```
  volatile type *p;
  ```

- Or:

  ```
  type volatile *p;
  ```

- Volatile pointer that points to non-volatile data:

  ```
  type* volatile p;
  ```

- Volatile data:

  ```
  volatile type data;
  ```

- Or:

  ```
  type volatile data;
  ```

As a word of caution, be advised that volatile is a characteristic of the C language, and it has nothing to do with CPU cache. So in addition to the volatile keyword, those memory-mapped addresses are supposed to be configured as "non-cacheable" in MMU. Otherwise, explicit cache invalidation/flush has to be used when these addresses are being accessed.

Type-Qualifier "const"

In addition to volatile, there is another type-qualifier that is often used in firmware coding: const. The const keyword will inform the compiler that certain storage locations are read-only, and any further attempts to write these locations should be flagged as error by the compiler. And similar to volatile, the actual position of const in the variable declaration could produce a different binary code, as shown in the following examples:

- Constant data:

  ```
  const type data = ...;
  ```

- Pointer that points to constant data:

  ```
  const type* p = &gVar;

  p = NULL;     // valid statement
  (*p) = ...;   // will produce compile error.
  ```

- Constant pointer that points to non-constant data:

  ```
  type* const p = &gVar;

  p = NULL;     // will produce compile error

  (*p) = ...; // valid statement
  ```

Access Peripheral Registers

In hardware, peripheral registers are usually mapped into a memory space. The following mixed use of const and volatile is recommend when these register are being operated on:

- Use volatile unsigned int * const to specify register address: In practice, unsigned int can also be replaced by U32 or U16. Here's an example of such a practice, where control register is written, followed by waiting on the busy flag.

  ```
  #define BUSY_FLAG (1 << 0)
    ...

      volatile U32* const REG_CONTROL = (U32*) 0xABCD0000;
      volatile U32* const REG_STATUS  = (U32*) 0xABCD0004;

      (*REG_CONTROL) = ... ; // correct statement

      REG_CONTROL = ... ;   · // will produce compile error

      while((*REG_STATUS) & BUSY_FLAG); // wait on flag

    ...
  ```

- Use const volatile type * to explore a read-only data buffer: If the peripheral exposes a buffer of read-only data to the CPU, a pointer of const volatile type * can be used to explore the data, with an additional sanity check from the compiler to prevent inadvertent writes on the buffer.

  ```
  const volatile U32 *p;
  U32 data1, data2, data3;

  volatile U32* const DATA_BUFFER = (U32*) 0xABCD0008;

  p = DATA_BUFFER;

  data1 = *p++;
  data2 = *p++;
  data3 = *p++;

  (*p) = ...;   // will produce compile error
  ```

- Use volatile type * to read/write the data buffer: If the peripheral exposes a bi-directional buffer to the CPU, a pointer of volatile type * can be used to explore the data.

  ```
  #define DATA_BUFFER 0xABCD0008

  volatile U32 *p;
  U32 data;
  ```

```
p = (U32*)DATA_BUFFER;

*p++ = ...;   // write data
data = *p;    // read data
```

Atomic Operation and Critical Section

As stressed in previous sections, it is quite possible for data to be modified externally, and that is the main reason why the volatile keyword was introduced in C. In addition to the recommended practice, firmware engineers should also recognize the atomic operation.

For example, if p is a pointer to 32-bit volatile data (memory-mapped to peripheral data buffer), the statement (*p) = ...; will generate bus-write transactions. In a 32-bit system, such a statement will be compiled into one instruction, which corresponds to one bus-write transaction, so the statement by itself is an atomic operation. However, in a 16-bit system, an atomic bus-write operation can only take 16-bit data, and such a statement will be chopped into two consecutive writes on the bus. However, if an interrupt is also enabled when these transactions are being carried out, chances are that an ISR (Interrupt Service Routine) could cut in between those two atomic bus writes. And even worse, the ISR may operate on the same peripheral data buffer as well, which generates a corner case that has never been mapped out in the first place.

In general, if it is possible that more than one agent will operate on the same object concurrently, and if such an operation is not atomic by itself, some protection measures have to be taken to avoid conflicts. And the piece of code that operates on the object is often called *critical section*[3]. There are many ways to protect critical sections, such as disabling-interrupt (for single processor), spin-lock, semaphore, etc. Inquisitive readers can find more information in Refs [12][13] for critical section protection. C++ has built-in support for the atomic operation, which will be discussed in the next chapter.

Finite State Machine

In the eyes of control theory, a system can be modeled by employing state variables. And the state of a system is a set of variables such that the knowledge of these variables and the input functions will, with the equations describing the dynamics, provide the future state and output of the system (Ref [9]). Since nearly every embedded system contains some sort of control logic, the state variable model can be applied universally, and this leads to one thing—the implementation of FSM (Finite State Machine).

As exemplified in Figure 6-2, FSM can be represented by state diagram, and there are two major approaches for its implementation: the hardware approach and the software approach. In embedded systems, the hardware approach usually means FPGA/ASIC, and it will be discussed rigorously in later chapters. Conversely, the software approach chooses to realize FSM through a microprocessor, which is addressed here in this section.

---

[3]Refer to your computer science text book if you want to find a more pedantic definition of "critical section".

SIDE NOTES

The implementation choice between hardware (FPGA/ASIC) and software (microprocessor) is not confined to FSM only. From the perspective of design philosophy, any embedded system design is actually an art of compromising on multiple dimensions, including cost, functionality, power consumption, reliability, time-to-market, etc. The balance between hardware and software is just one of those compromises.

For example, an n-tap FIR filter can be implemented by FPGA with n multipliers in parallel, or it can be realized in DSP with n iterations of a MAC (Multiply-Accumulate) operation. Both approaches have pros and cons. FPGA approaches can be carried out with relatively low clock frequency, but they lack the flexibility that microprocessor can offer. And their overall contribution to BOM cost is hinged on various factors. That's why each embedded system is unique and the design choice has to be made on case-by-case basis.

To illustrate the software approach, I will start out by employing FSM to simulate a project lifecycle. As far as project management is concerned, each embedded system project will go through five phases in its lifecycle: initiating, planning, controlling, executing, and closing (Ref [10]). The transition among these five phases can be simulated by using a FSM with random input, as illustrated by the state diagram in Figure 6-2. (The inputs in Figure 6-2 are just some random numbers I've generated for the sake of discussion. Don't get hung up on their meanings.)

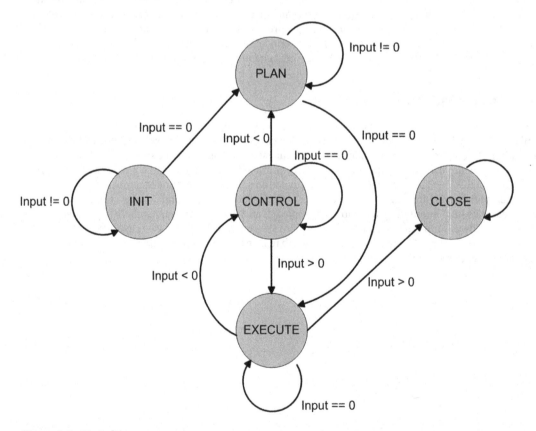

Figure 6-2. State diagram

To implement a FSM in C, two things are necessary: the storage space for state variable and the code to determine the state transition and output. One way to achieve these goals is to use a big switch statement, as demonstrated in Listing 6-15.

Listing 6-15. FSM in the C Language

```c
enum FSM_State {
    STATE_INIT,
    STATE_PLAN,
    STATE_CONTROL,
    STATE_EXECUTE,
    STATE_CLOSE
};

char *state_name_string[] = {
    "STATE_INIT",
    "STATE_PLAN",
    "STATE_CONTROL",
    "STATE_EXECUTE",
    "STATE_CLOSE"
};

enum FSM_State FSM(int input)
{
    static enum FSM_State state = STATE_INIT;
    enum FSM_State prev_state;

    printf ("current state = %s\n", state_name_string[state]);

    prev_state = state;

    switch (state) {
        case STATE_INIT:
            logo_print();
            if (input == 0) {
                state = STATE_PLAN;
            }

            break;

        case STATE_PLAN:

            if (input == 0) {
                printf ("proceed to execute\n");
                state = STATE_EXECUTE;
            }

            break;
```

```
    case STATE_EXECUTE:

        if (input > 0) {
            printf ("about to close\n");
            state = STATE_CLOSE;
        } else if (input < 0) {
            state = STATE_CONTROL;
        }

        break;

    case STATE_CONTROL:
        if (input > 0) {
            state = STATE_EXECUTE;
        } else if (input < 0) {
            state = STATE_PLAN;
        }

        break;

    case STATE_CLOSE:
        printf ("Game Over!\n");
        printf ("ABOUT_CRC32 = 0x%x\n", ABOUT_CRC32);
        break;

    default:
        printf (" unknown state, index = %d\n", state);
        break;
    } // End of switch

    return prev_state;
} // End of FSM()
```

The FSM() function in Listing 6-15 forms the engine of finite state machine, and it should be called every time a new input is available. FSM state is stored in a static variable in Listing 6-15. You can also make the FSM state a parameter of the FSM() function and keep track of it outside.

The state transition and output will be determined by each branch of the switch statement. The plus side of the switch statement is that it offers great flexibility. However, if the state diagram grows, so does the switch statement, which will make the code cluttered and bloated if the numbers of states is really big.

To keep things clean and compact in Listing 6-15, you can consolidate the code in each branch into functions. Or you can take a different approach, as suggested in Ref [11], by building the FSM with three tables:

- Have a state-table to describe all the states in the FSM. In this case, it will be five entries for STATE_INIT, STATE_PLAN, STATE_CONTROL, STATE_EXECUTE, and STATE_CLOSE.

- Have an event table to describe all the possible input. In this case, it will be EVENT_POSITIVE (input > 0), EVENT_ZERO (input == 0), and EVENT_NEGATIVE (input < 0).

- Having a state-transition table to describe state transitions and corresponding event handler functions. This table is actually a two-dimensional array, with each element being a structure of the next state and handler function, as exemplified in Table 6-1. (The handler function can be dummy() for no action or msg_print() for message printing.)

Table 6-1. *State Transition Table*

	EVENT_POSITIVE	EVENT_ZERO	EVENT_NEGATIVE
STATE_INIT	STATE_INIT, logo_print()	STATE_PLAN, logo_print()	STATE_INIT, logo_print()
STATE_PLAN	STATE_PLAN, dummy()	STATE_EXECUTE, msg_print()	STATE_PLAN, dummy()
STATE_CONTROL	STATE_EXECUTE, dummy()	STATE_CONTROL, dummy()	STATE_PLAN, dummy()
STATE_EXECUTE	STATE_CLOSE, msg_print()	STATE_EXECUTE, dummy()	STATE_CONTROL, dummy()
STATE_CLOSE	STATE_CLOSE, msg_print()	STATE_CLOSE, msg_print()	STATE_CLOSE, msg_print()

If you have a very large state machine, this is a systematic approach to keep things tight and clean. For those inquisitive minds, refer to Ref [11] for all the coding details.

Last but not least, semaphores can be adopted in Listing 6-15 if it is going to be used in a multithread setting. Instead of sharing the same static state variable across multiple threads, you might also consider maintaining separate state variable for each thread, and passing it as a new function parameter to FSM().

Firmware Version and About Signature

For firmware team of any size, be it a one-man show or a choral group, there ought to be some sort of version control process in place to track the firmware change over time. From the standpoint of building process, it means the version number should be stamped into the binary image when the firmware is being built. The version number is usually composed of two parts[4]: the major version and the minor version. Depending on the version control process, the minor version sometimes can be a mixture of numbers and letters, while the major version is made up only of numbers. The whole version number is often displayed by concatenating the major version and minor version with a dot, such as 11.0, 10.20c, etc.

To build the version number into firmware image, the traditional approach is to define two macros (VMAJOR and VMINOR) in a makefile and insert them into a constant string in firmware. That constant string is often called the "about signature," and it also includes product name, company name, and copyright claim, just like what you see in the About menu of most applications.

The version number should be treated as a string since the minor version might include letters. That is to say, you need to put quotation marks on VMAJOR and VMINOR when they are referenced in the C code. In C, this can be achieved by using the # in macro definition, as illustrated in Listing 6-16.

---

[4]Sometimes people also put the build number or OEM ID as the third part of the version number.

Listing 6-16. Version Number Added to the About Signature

```
In Makefile, define the version number (such as 1.3g) in VMAJOR and VMINOR
CFLAGS += -DVMAJOR=1 -DVMINOR=3g
```

```
In C File, put quotation marks on VMAJOR and VMINOR

#define VERSION_STR(s) VERSION_STR_(s)
#define VERSION_STR_(s) #s

#define ABOUT_SIGNATURE_LENGTH      512

const U8 about_signature[ABOUT_SIGNATURE_LENGTH] = \
  "\n================================================================\n"\
  VERSION_STR(PRODUCT_NAME)"\n"\
  "Version "VERSION_STR(VMAJOR)"."VERSION_STR(VMINOR)"\n"\
  "Copyright (c) 2099 PulseRain Technology (www.pulserain.com).\n"\
  "All Rights Reserved.\n"\
  "================================================================\n\n";
```

Speaking of the "about signature", be advised *not* to build timestamps into the firmware image. The reason is that you might want to generate the same binary image given the same source code and same building tools and follow the same building process. This one-to-one mapping between source code snapshot and binary image is crucial during regression testing. If you can reproduce exactly the same binary image from the same source code no matter when you build it, you can be confident that your version control process is doing its job properly.

Embedding the "about signature" in a binary image can protect your intellectual property. The "about signature" ought to be the first message sent to the console, and it serves as a strong proof of authorship. However, the world is never short of crackers and copycats. IP thefts could rip you off by using hex-editors to modify the "about signature," and then plagiarize your binary image to their copycat hardware.

One thing you can do to protect yourself is to verify the integrity of the "about signature" through CRC checksum and navigate the code flow into an endless loop if "about signature" were compromised, as demonstrated in Listing 6-17. Of course, crackers can always find other ways to circumvent this, but CRC checksum makes their job harder.

Listing 6-17. CRC Checksum of About Signature

In Makefile, specify the correct CRC32 checksum in ABOUT_CRC32 macro

```
CFLAGS += -DABOUT_CRC32=0x68DAC604
```

In C File, check the CRC32 checksum of "about signature"

```
int main()
{
    while(CRC32(about_signature, ABOUT_SIGNATURE_LENGTH)!= ABOUT_CRC32);

    CONSOLE_PRINT("%s\n", about_signature);

    ...

} // End of main()
```

Summary

This chapter contained an overview of various programming languages, which concluded that C and C++ are the two preferred choices for embedded systems. After that overview, this chapter started with those pitfalls and fallacies when C is applied to embedded systems. A few coding tricks to ensure reliability were thus introduced. In addition, some routine jobs, like Finite State Machine, code version, and code integrity protection are also examined with practical templates.

The next chapter discusses C++ and script languages.

References

1. C++ for Kernel Mode Drivers: Pros and Cons, WinHEC 2004 Version, Microsoft Corporation, April 10, 2004
2. "Why don't we rewrite the Linux kernel in C++?" (The Linux kernel mailing list FAQ http://www.tux.org/lkml/#s15-3) or (http://harmful.cat-v.org/software/c++/linus)
3. Object-Orientated Programming with ANSI-C, Axel-Tobias Schreiner, December, 1994
4. Portable Inheritance and Polymorphism in C, by Miro Samek, *Embedded Systems Programming,* December, 1997
5. Porting Your Driver to 64-Bit Windows, MSDN Library, September 7, 2011
6. Catching errors early with compile-time assertions, by Dan Saks, *Embedded Systems Programming,* July, 2005
7. Intro to GNU Assembly Language on Intel Processors, Prof. Godfrey C. Muganda, North Central College, February 29, 2004
8. Place Volatile Accurately, by Dan Saks, *Embedded Systems Programming,* November, 2005
9. *Modern Control Systems (10th Edition).* Richard C. Dorf and Robert H. Bishop, Pearson Education, Inc., 2005
10. The Principles of Project Management, Meri Williams, SitePoint Pty. Ltd., February, 2008
11. The Embedded Finite State Machine. Bo Berry, *Dr. Dobb's Journal,* May 13, 2009
12. *UNIX SYSTEMS Programming, Communication, Concurrency, and Threads.* Kay A. Robbins, & Steven Robbins, Prentice Hall PTR, 2003
13. *Modern Operating Systems (3rd Edition).* Andrew S. Tanenbaum, Pearson Education, December, 2007

■ ■ ■

## CHAPTER 7

Firmware Coding in the C++ and Script Languages

There are only two kinds of languages: the ones people complain about and the ones nobody uses.

—Bjarne Stroustrup, the Father of the C++ Language

To continue where we left off, this chapter demonstrates how C++ can be applied to embedded systems. Script languages are also examined in this chapter.

Firmware Coding in C++

C++ is a powerful language, and it offers a lot of cool features to embedded system developers. However, there are a few things you should keep in mind before you jump onto the C++ wagon:

- Unlike C, the standard of C++ keeps changing. Sometimes it is hard to find a compiler that can support all the latest features. And all bets are off if you can't find a good C++ cross compiler for your microprocessor.

- C++ is far more complicated than C. And its steep learning curve has to be taken into account for staff hiring and training, as well as the project planning.

- The C++ code in this book was compiled and tested under g++ 4.9.2, with the -std=c++11 option on.

- Note that C++ is a very powerful language, and inquisitive readers are strongly recommended to peruse Ref [12][13] to get an idea of its full power. The purpose of this chapter is to demonstrate the practicality for coding firmware in C++.

- This chapter is broken into two parts. The first part discusses those C++ features that are mostly seen as enhancements to C. The second part examines the new concepts that C++ offers and their relevance to embedded system development.

Enhancements to the C Language

As shown in the previous chapter, there are tricks you can play in C, such as compile-time assertion, to improve coding quality. Many of those tricks have been absorbed into C++ as part of the language standards.

Explicit Type of Bit-Width

As mentioned, it is recommended to explicitly associate a data type with its bit-width. For C, it is done through typedef. For C++, the C++ standard library has built-in type definitions like std::uint16_t, std::uint32_t, and so on, and these come in handy for explicit bit-widths. All you have to do is #include <cstdint>.

Use "constexpr" to replace "define"

In the previous chapter, the register address was defined as volatile U32* const in C. Although this method is still supported by C++, an alternative offered by C++11 is to use constexpr. constexpr was introduced in C++11 to replace the traditional macro definitions for compile-time constants. Listing 7-1 is an example of defining register addresses and bit masks using constexpr. And as illustrated in Listing 7-1, reinterpret_cast and volatile should both be used when accessing registers defined by constexpr.

Listing 7-1. Using constexpr for Register Definition

```
class ABC
{
    public :

    //=======================================================================
    // Register Definition
    //=======================================================================

    //++++++++++++++++++++++++++++++++++++++++++++++++++++++++++++++++++++++++
        static constexpr uint32_t REG_A = 0x0abc;
            static   constexpr uint32_t BIT_XXX_ENABLE  = (0 << 8);
            static   constexpr uint32_t BIT_YYY_DISABLE = (1 << 8);

    ...

};

//==================================================
// Access register using reinterpret_cast
//
// *reinterpret_cast<volatile uint32_t*>(ABC::REG_A) = 0xABCD1234;
```

The constexpr function is capable of doing very complicated math as long as the result is determinable at compile-time. Another restriction of constexpr is that its body has to be a return statement. Notwithstanding this drawback, constexpr can be a handy way to generate lookup tables automatically (see Listing 7-2).

Listing 7-2. Using the constexpr Function for a Lookup Table

```
constexpr uint16_t cos_int (int i)
{
   return round(cos (static_cast <long double>(i) / 2048 * 3.14159 / 4) * 32767);
}

const std::array<uint16_t, 2048> cos_table{{
   cos_int(0),
   cos_int(1),
   cos_int(2),
   ...
   cos_int(2046),
   cos_int(2047)
}};
```

As demonstrated in Listing 7-2, a cos table of 2048 items are generated with constexpr. Since cos_int is crunched at compile-time, floating-point math can be avoided completely with a lookup table.

The drawback of Listing 7-2 is that its lookup table has to be manually (or with the assistance of script) populated when the table size increases. It can be quite cumbersome when the table size is big. This is mainly caused by the restriction that constexpr function can only have return state as its function body.

To break free from this restriction, you can take on template specialization and class inheritance, as illustrated in Listing 7-3.

Listing 7-3. Generating a Lookup Table Automatically

```
#include <cmath>
#include <array>

template<typename T>
constexpr T look_up_table_elem (int i)
{
   return {};
}

template<>
constexpr uint16_t look_up_table_elem (int i)
{
   return round(cos (static_cast <long double>(i) / 2048 * 3.14159 / 4) *
   32767);
}

template<typename T, int... N>
struct lookup_table_expand{};

template<typename T, int... N>
struct lookup_table_expand<T, 1, N...>
{
    static constexpr std::array<T, sizeof...(N) + 1> values =
{{look_up_table_elem<T>(0), N... }};
};
```

```
template<typename T, int L, int,.. N> struct lookup_table_expand<T, L, N...>
: lookup_table_expand<T, L-1, look_up_table_elem<T>(L-1), N...> {};

template<typename T, int... N>
constexpr std::array<T, sizeof...(N) + 1> lookup_table_expand<T, 1,
N...>::values;

const std::array<uint16_t, 2048> lookup_table =
lookup_table_expand<uint16_t, 2048>::values;
```

Listing 7-3 is chiefly inspired by the ideas presented in Refs [14][15]. Among all the functions presented in Listing 7-3, look_up_table_elem is the one that generates the value for each table item. And the structure lookup_table_expand is template-specialized and inherited iteratively. typename T is introduced here to make Listing 7-3 suitable for types other than int.

However, Listing 7-3 has its own restrictions imposed by compilers. Since it adopts the idea of iterative inheritance, its depth of iteration is tied to the size of the lookup table. Under a GNU C++ compiler, a message like "template instantiation depth exceeds maximum of xxx" will pop up when the lookup table is too big. Fortunately, a GNU C++ compiler allows users to bump up the template instantiation depth with the -ftemplate-depth option.

If adjusting the compiler option bothers you, there is another way to do this by divide-and-conquer, which can lower the iteration depth to log_2^N instead of N. Inquisitive readers can find more information in Ref [19].

Align the Data Structure

As mentioned in the previous chapter, for embedded systems, it is critical to align data structures to a memory boundary. In C, alignment is dealt with by using compiler attributes, as well as macro definitions. For C++11, the following built-in features can be used:

- alignof operator: alignof acts the same way as the TYPE_ALIGNMENT macro defined in the previous chapter. But unlike the macro definition used by C, alignof is a built-in operator with native support for C++11. Listing 7-4 shows an example of it.

Listing 7-4. The alignof Operator

```
#define PACKED __attribute__((packed))

typedef struct {
    uint8_t  a;
    uint32_t b;
    uint16_t c;
} PACKED STRUCT_PACKED;

typedef struct {
    uint8_t  a;
    uint32_t b;
    uint16_t c;
}STRUCT_NOT_PACKED;

static_assert(alignof(STRUCT_PACKED) == 1, "");
static_assert(alignof(STRUCT_NOT_PACKED) == 4, "");
```

- alignas specifier: In addition to the `alignof` operator, C++11 also provides the `alignas` specifier to place data structure and variables on memory boundaries, as shown in Listing 7-5.

Listing 7-5. The alignas Specifier

```
typedef struct {
    uint8_t  a;
    uint32_t b;
    uint16_t c;
}STRUCT_NOT_PACKED;

typedef struct alignas(128) {
    uint8_t  a;
    uint32_t b;
    uint16_t c;
}STRUCT_ALIGNED;

alignas(128) STRUCT_NOT_PACKED abc;

static_assert(alignof(STRUCT_ALIGNED) == 128, "");

int main()
{
   std::cout << "abc is aligned to " << alignof(abc);
}

//=============================================================
// Output:
//
//   abc is aligned to 128
//
```

- Standard library support: In addition to the native language support, the C++ standard library also provides functions to assist memory alignment. One of the `std` functions is `std::align`, which can help allocate the memory buffer while aligning buffer pointers to a specified boundary. Inquisitive readers are suggested to go through the standard library manual for more detailed information.

Compile Time Assertion

C does not have native support for compile-time assertion. And in the previous chapter, the macro `C_ASSERT` was defined to remedy this issue. Fortunately, C++11 has created the `static_assert` keyword to support compile-time assertion natively. Examples can be found in Listings 7-4 and 7-5.

Use "nullptr" to replace "NULL"

In C, an invalid pointer address is defined as NULL, which is actually a constant value of zero. For C++, things become more complicated due to type deduction and function/template overload. It is thus recommended to use `nullptr` instead of NULL for invalid address whenever possible. The `nullptr` carries type information and can be less error prone.

133

Function Object and Lambda Expression

It is not uncommon for embedded systems to execute certain functions based on a mode (or other parameters/states) determined at runtime. For C, this can be implemented with a function pointer, as illustrated in Listing 7-6.

Listing 7-6. A Function Pointer

```c
#include <stdio.h>

typedef int (*T_FUNC_POINTER)(int, int);

int func_add (int a, int b)
{
    return a + b;
}

int func_sub (int a, int b)
{
    return a - b;
}

int func_add_sub (int mode, int a, int b)
{
    T_FUNC_POINTER func_table[2]={func_add, func_sub};
    return (*(func_table[mode]))(a, b);
}

int main()
{
    printf ("add = %d\n", func_add_sub(0, 6, 4));
    printf ("sub = %d\n", func_add_sub(1, 6, 4));
}
```

However, a function pointer is nothing but an address to the function entry. It cannot carry any state information. An alternative offered by C++ is to use a function object (or *functor*), as shown in Listing 7-7.

Listing 7-7. Function Object and Lambda Expression

```cpp
class func_obj_add_sub
{
    public:
            func_obj_add_sub (int mode): _mode {mode} {};

            int operator() (const int a, const int b)
            {
                    if (_mode == 0) {
                            _mode = 1;
                            return a + b;
                    } else {
                            _mode = 0;
                            return a - b;
                    }
            }
```

```
    private:
            int _mode;
};

int main()
{
    auto add = [](const int a, const int b){return a + b;};
    auto sub = [](const int a, const int b){return a - b;};

    func_obj_add_sub add_sub{0};

    std::cout << "1st call " << add_sub (6, 4) << "\n";
    std::cout << "2nd call " << add_sub (6, 4) << "\n";
    std::cout << add (6, 4) << " " << sub (6, 4) << "\n";

}
```

Function objects (functors) can be seen as classes with the () operator overridden. Because they are essentially classes, they can carry far more information than function pointers. As demonstrated in Listing 7-6, the class func_obj_add_sub overrides the () operator, and the () function will change its internal state (the private variable _mode) every time it is invoked. Thus, it is recommended to replace the function pointer with function objects for the interest of flexibility.

Lambda expressions can be used to generate an anonymous function object, and they come in handy when the function object is lightweight. Examples can be found in Listing 7-6. It is also a terse way to generate callback functions. Inquisitive readers can find more information in Ref [12].

Atomic Operation

Atomic operations were mentioned in Chapter 6. C++11 now offers support for atomic operation in its standard library. Listing 7-8 shows an example of it. Inquisitive readers are suggested to go through Ref [12] for more information.

Listing 7-8. Atomic Operation

```
#include <iostream>
#include <atomic>

int main()
{
    volatile std::atomic<uint16_t> aaa {1234};

    uint16_t bbb;

    bbb = aaa.load();
    std::cout << "bbb = " << bbb << "\n";

    aaa.store(2346); bbb = aaa.load();
    std::cout << "bbb again = " << bbb << "\n";
}
```

New Concepts Offered by C++

C++ is not merely an enhancement to C. It also brings new concepts and new methodologies to the table.

RAII (Resource Acquisition Is Initialization)

In my personal opinion, RAII is probably the most valuable feature that C++ brings to the table. Resource management has always been a big headache for C as its resource acquisition and release are all conducted manually. And the resource mismanagement by unwary programmers is often the source of memory leakage or deadlock. The remedy that C++ provides is tying the resource management to the lifespan of an object. During the creation of that object, its constructor is called and the resource is acquired. When the object's lifespan ends, its destructor is invoked (implicitly by C++) and the resource is released there (see Listing 7-9).

Listing 7-9. RAII (Resource Acquisition Is Initialization)

```
{
   // object creation, constructor called
   class_name obj {initializer_list};

   //
   // do something
   //

} // obj lifespan ends, its destructor is called
```

The merit of RAII is that resource acquisition and release are both contained in the object itself. There is no need to scatter them (acquisition/release) all over the place. Listing 7-9 shows an example of it. Note that the destructor does not have to be called explicitly by the programmers. The C++'s built-in mechanism will invoke the destructor when the execution flow moves out of the closing brace (where the lifespan of the object ends).

For dynamically allocated memory buffers, programmers can still acquire the buffers in the constructor and release them in the destructor. However, for that purpose, C++11 has provided the smart pointer based on RAII principle in its standard library. The two prominent smart pointer types are `std::unique_ptr` and `std::shared_ptr`. Inquisitive readers are suggested to go through the standard library part in Ref [12] for more information on smart pointer.

Speaking of dynamically allocated memory, I have a few words to add here. As mentioned, for embedded systems, especially those working close to the bare bones hardware, you should try to avoid C standard library as much as you can. This is also true for the C++ standard library. However, I have to admit that the C++ library offers richer features than its C counterpart. If you choose the C++ standard library due to its handy features, you should make sure that the part you are using does not allocate memory behind your back.

One example is the `std::vector` template class. The vector will grow when more elements are pushed into it. Obviously, the extra space it needs to grow comes with dynamic allocation. For that, you have two choices:

- Use a different data structure with a fixed size, such as `std::array`.

- If you can't avoid using `std::vector` completely, at least use placement new to do allocation from a predetermined buffer pool. In this way, its memory range can be more controllable.

As mentioned, data structures with dynamic memory allocation (new/delete or malloc/free) should be avoided as much as possible, especially in ISR or DMA where the function call cannot be blocked, or the memory pool has to be non-paged. It is better to manage the memory buffer on your own in those circumstances.

Template and Overloading

Template and function/operator overloading are great ways to make terse code. And they are the building blocks to fulfill the lofty ideals of *generic programming* and *metaprogramming*.

Generic Programming

Notwithstanding the academic pedantry, generic programming can be seen practically as programming without specifying the data type. The actual data type is specified later when the template is instantiated and determined at compile time.

Here is an example as how generic programming can make code terse. Assume you would like to implement a function to calculate $-x \cdot y$ for both integers and complex numbers. In C, you usually have to write two functions to deal with each type separately, as illustrated in Listing 7-10. The functions mult_neg_int() and mult_neg_complex() are written to deal with the integer and the complex type, respectively.

Writing the same (or very similar) algorithm for each data type could be a very tedious job. And if you want to modify the algorithm in the future, the modification has to be applied to each and every data type and function that is involved, which by itself is an error-prone process.

Listing 7-10. Separate Function for Each Type

```
typedef struct {
    int x;
    int y;
}MY_COMPLEX;

int mult_neg_int (int x, int y)
{
    return x * y * (-1);
}

MY_COMPLEX mult_neg_complex (MY_COMPLEX *pa, MY_COMPLEX *pb)
{
    MY_COMPLEX tmp;
    tmp.x = (pa->y) * (pb->y) - (pa->x) * (pb->x);
    tmp.y = -((pa->x) * (pb->y) + (pa->y) * (pb->x));

    return tmp;
}
```

And with the help of template and operator overloading, C++ could take the following approaches, as shown in Listing 7-11.

Listing 7-11. Function Template and Operator Overloading

```cpp
#include <iostream>

class my_complex
{
    public:

        explicit my_complex(int x) : _x {x}, _y {0}
        {
        }

        explicit my_complex(int x, int y) : _x {x}, _y {y}
        {
        }

        friend my_complex operator*(const my_complex &a,
                                    const my_complex &b)
        {
            int x, y;

            x = a._x * b._x - a._y * b._y;
            y = a._x * b._y + a._y * b._x;

            my_complex tmp{x, y};

            return tmp;
        }

        friend std::ostream& operator<<(std::ostream& os,
                                        const my_complex& a)
        {
            return os << "(" << a._x <<", " << a._y <<")\n";
        }

    private :
        int _x, _y;
};

template <typename T>
T mult_neg (const T& a, const T& b)
{
    T minus_one{-1};

    return (a * b * minus_one);
}

int main()
{
```

```
my_complex aaa{1,2};
my_complex bbb{3,7};

int t = mult_neg(2,3);
my_complex ccc = mult_neg(aaa, bbb);
//my_complex ddd = mult_neg(bbb, aaa);

std::cout << t << "\n";
std::cout << ccc << "\n";
}
```

In Listing 7-11, only one function mult_neg() is used to implement the algorithm for calculating $-x \cdot y$, and it works on an abstract type T. When it comes to the complex type my_complex, the trick is all in the constructor and operator overloading. Class my_complex defines two constructors and one of them takes a scalar value and makes a real number out of it. In this way, my_complex minus_one{-1} can be properly initialized into (-1, 0). On the other hand, the * operator is overloaded to support complex number multiplication. So (a * b * minus_one) is actually calling the overloaded function when mult_neg() is instantiated with the my_complex type. And to make the display easy, the << operator is also overloaded in Listing 7-11.

Note that generic programming is just a mechanism to automatically populate the code for each data type. Behind the scenes, those data types still have their own versions of implementation, as demonstrated in Lab 7-1.

▪ **Labs**   To ease the job of comparison, iostreams and the << operator overloading are removed from Listing 7-12, as shown next (filename: mult_neg.cpp).

Listing 7-12.   Test Program: mult_neg.cpp

```
class my_complex
{
    public:
        explicit my_complex(int x) : _x {x}, _y {0} {}
        explicit my_complex(int x, int y) : _x {x}, _y {y} {}

        friend my_complex operator*(const my_complex &a,
                                    const my_complex &b)
        {
            int x, y;

            x = a._x * b._x - a._y * b._y;
            y = a._x * b._y + a._y * b._x;

            my_complex tmp{x, y};
            return tmp;
        }
```

```cpp
    private:
        int _x, _y;

};

template <typename T>
T mult_neg (const T& a, const T& b)
{
    T minus_one{-1};

    return (a * b * minus_one);
}

int main()
{
    my_complex aaa{1,2};
    my_complex bbb{3,7};

    //int t = mult_neg(2,3);
    my_complex ccc = mult_neg(aaa, bbb);
    my_complex ddd = mult_neg(bbb, aaa);
}
```

LAB 7-1: TEMPLATE INSTANTIATION

If you compile mult_neg.cpp with the -O0 optimization:

```
$> g++ -std=c++11 -O0 -c mult_neg.cpp
$>
```

And its symbol list will be as following:

```
$> nm mult_neg.o |c++filt


    ...
        U __main
00000000 T my_complex mult_neg<my_complex>(my_complex const&,
                                my_complex const&)
00000000 T operator*(my_complex const&, my_complex const&)
00000000 T my_complex::my_complex(int)
00000000 T my_complex::my_complex(int, int)
00000000 T _main
```

Notice that only mult_neg<my_complex> appears in the symbol list.

Now if you uncomment the line of int t = mult_neg(2,3); and comment out the last line of my_complex ddd = ..., you will get a bigger object file and a symbol list as follows:

```
...
         U ___main
00000000 T my_complex mult_neg<my_complex>(my_complex const&,
                                    my_complex const&)
00000000 T int mult_neg<int>(int const&, int const&)
00000000 T operator*(my_complex const&, my_complex const&)
00000000 T my_complex::my_complex(int)
00000000 T my_complex::my_complex(int, int)
00000000 T _main
```

As you can see, we've replaced a complex operation with an integer operation, but the code size has increased instead of decreasing. The reason is that the integer operation has caused the instantiation of the mult_neg<int> function, which does not exist in the previous run.

In other words, there are still two functions for each data type in the binary code, just like the case in Listing 7-10 for C. But C++ does this automatically with template instantiation and operator overloading.

Also, note that such automation comes with a price. The C++ code is usually bigger than its C counterpart in this regard. Personally I think the terseness is worth more than the overhead as long as it meets the real-time requirement and the size constraint.

Metaprogramming

For all practical purposes, metaprogramming in C++ can be seen as generating code at compile time using templates primarily. Listing 7-3 is a good example of it, in which multiple structures are generated through template expansion.

However, metaprogramming can lead to inscrutable code that few can understand. So my two cents is to use template metaprogramming only to the point where it is comprehensible. Keep an eye on the image size to prevent bloated code.

Error Handling

To err is human. Error handling is an important part of embedded system development. As compared to C, C++ offers the "exception" mechanism by handling errors in throw and catch, which is a terse way to deal with errors.

However, for firmware code that works closely to the bare bones hardware, predictability and reliability are far more important than terseness. In that sense, it is often discouraged to apply C++'s exception mechanism to embedded system for error handling. Not only will C++ exception increase the code size, it will also make the error handling unpredictable since it is hard to determine which catch will get the throw without going through the whole piece of code. And spilling error handling all over is not a good idea for embedded systems. This is especially true when you are working on very low-level code (i.e., code close to the hardware).

So instead of using exceptions, the traditional method is preferred. That is to say, the error should be handled *based on return codes and other old-fashioned and tedious, but predictable, techniques* (Ref [16]).

OOP (Encapsulation, Polymorphism, and Inheritance)

OOP (Object Oriented Programming) is a great idea that facilitates clean and robust code. OOP mainly comprises three fundamental concepts—Encapsulation, Polymorphism, and Inheritance—and they are reflected in C++'s various language features. However, note that the idea of OOP is not confined to specific languages. C can also borrow a page from the OOP concept, as demonstrated in Ref [20][21]. And when it comes to the software architecture design (embedded system included), OO can definitely provide some useful clue trails.

Boost Library

The Boost C++ Library (http://www.boost.org) can be a good addition to the C++ standard library, and its license term is pretty open and lenient. Thus embedded systems should see it the same way as the C++ standard library.

In addition, the Boost Library comes with its own code build system—BJAM—which will be discussed in the next chapter.

C/C++ Mixed Programming

Since C is the English language de facto for embedded systems, mixed language programming would be inevitable if C++ were added to the picture.

Between C and C++, the interaction is mostly through function calls. However, the calling conventions of C and C++ are different[1]. To enable C and C++ to talk to each other, the functions and C head files must be declared with extern "C", as shown in Listing 7-13.

Listing 7-13. Using extern "C" for the C Language Calling Convention

```
extern "C" {
    #include "stdio.h"
}

extern "C" void read_ISR (void* isr_context);
```

On the other hand, C++ functions should have a C wrapper with the extern "C" modifier as well; this even applies to the main() function, as demonstrated in Listing 7-14 for an example of main.cpp. In Listing 7-14, the LCD_16207 is a C++ class, and it is invoked within the main() function. The main() function also has a C calling convention in order to link properly.

---

[1]For all practical purposes, this calling convention can be seen as the way that the stack frame is formed, which includes the order that parameters are pushed onto the stack (from left to right or from right to left). There are two kinds of calling conventions: stdcall and cdecl. The default calling convention for your C and C++ compilers might be different.

Listing 7-14. Example of main.cpp

```cpp
#include <cstdint>
#include "LCD_16207.h"

LCD_16207 LCD_16207_i {};

extern "C" int main ()
{

    LCD_16207_i.display("ABC\n");
    //LCD_16207_i.clear();

    LCD_16207_i.move_cursor (1, 2);
    LCD_16207_i.clear_to_line_end();
    LCD_16207_i.display("Hello!");

    ...

    return 0;

}
```

Script Languages Overview

To ensure reliability, automated testing is an indispensable part of embedded system development. And thanks to their flexibilities, script languages come in handy for this job. In fact, today's script languages are so powerful that they can even be used as the main driving force under certain circumstances.

Compiled Versus Script Languages

As mentioned, C will be the first choice for writing firmware on embedded processors. However, to support firmware coding, some groundwork has to be laid a priori on the host side, which normally involves the following:

1. Pre-processing data. For example, to speed things up at the expense of space, you could generate look-up tables beforehand and include them in the firmware source code to save the precious CPU cycles.

2. Post-processing. For example, if things went south in the lab, you might end up with tons of data samples collected by the target processor. Most likely, they will be in some primitive form due to the nature of embedded systems. You have to whip up some customized code to analyze them on the host side.

3. Often, you will find yourself import/export data between different tools. That usually translates into processing text and matching string patterns.

4. For test automation, you might wish to have an intelligent test program running on the host side, reading the input from the console (or network, USB port, etc.) and acting accordingly, just like a human being would do. That test program is intended to run over a very long period of time to exert stress on the target.

5. For all these things, you might also want to quickly develop a user-friendly GUI.

As you can see, those jobs are just intermediate steps to support your firmware development and testing. You don't want them turn into a big sideshow to distract you from your main goal. Although those jobs can be done in any programming languages—such as C/C++/Java—a more practical approach is to do them in a script language because:

- Those jobs don't have to be done in real time as long as they can be completed in a reasonable time frame. Given the high clock frequency and multi-core architecture that PC can offer nowadays, compiled language wins very little over script language in this regard.

- Script language is rich in feature set of data structures, such as regular expression, string, list, associative array, etc., and it might take a significant amount of effort to implement the same feature in C/C++/Java from the ground up. Thus script language is largely favored in regard to test automation, numeric computation, and text processing.

- Most main-stream script languages support Tk toolkit, with which simple GUIs can be constructed very quickly. Since most script languages are cross-platform, so is the GUI they create.

Tcl/Tk, Expect, Perl, and Python

Despite the plethora of script languages, three stand out: Tcl, Perl, and Python (Expect is based on Tcl).

Tcl is a string-based command language (Ref [17]). As its mantra goes: "Everything is a string". Since its core mechanism is just string and string substitution, it brings both simplicity and oddness. One thing noticeably different from other languages is that in a Tcl procedure or block statement, its open curly brace has to be in the same line as the leading keyword. Otherwise, you have to put a backslash at the end as a line wrapper, as demonstrated in Listing 7-15. Obviously, this is a side effect caused by the "Everything is a string" mentality.

Personally, I found Tcl to be the hardest to master among these three script languages, thanks to its string-based mechanism. However, it offers two things that the other script languages are short of:

- Tcl is a very extensible language. You can add new commands to the Tcl shell by implementing them as an extension, or you can integrate the Tcl interpreter into your application and export new commands to it (Ref [17]). Such extensibility is greatly favored by EDA software that uses Tcl as the default script language.

- As one of its extensions, Tcl offers a graphic toolkit called Tk, which provides many basic GUI widgets, like buttons, textboxes, images, labels, etc. A graphical user interface can thus be whipped up without much effort. (To use Tk, you need to invoke your Tcl script with "wish", the graphic Tcl shell.) In this regard, the GUI provided by Perl and Python borrowed heavily from Tcl. (To communicate with Tk, Perl uses Perl/Tk and Python uses Tkinter.)

So my personal approach is to use Tcl/Tk only for GUI creation, and then delegate the real processing to other script languages, such as Perl or Python, and capture the in/out for display.

Listing 7-15. The Position of an Opening Curly Brace

```
proc Log {} { # opening curly brace on the same line
   ...

   if {$w == "left"} {
     button $f.pan_top.open_file -width 10 -text Load -command {Open_File}
   } else {
     button $f.pan_top.open_file -width 10 -text Save -command {Save_File}
   }
   ...
}
```

```
# Use back slash to continue at next line
proc Log {} \
{
   if {$w == "left"} \
   {
      button $f.pan_top.open_file -width 10 -text Load -command {Open_File}
   } else \
   {
      button $f.pan_top.open_file -width 10 -text Save -command {Save_File}

   }
}
```

When it comes to choosing between Perl and Python, there has been enough war of words between both camps. I'm not here to pass judgment, but just to offer personal opinions. My two cents:

- If you are doing text processing line by line, use Perl. Perl's regular expression is very powerful and comes with its own syntax.

- If you are dealing with binary files, use Python. Python could easily turn a binary file into a byte sequence and access its content randomly with sequence indexing.

- For number crunching, both of them do a good job. But I like Python better personally.

- As an analogy, Perl is to Python as CISC is to RISC. If you know your way around, Perl can be very compact. On the other hand, Python is more close to modern programming languages, with OOP in its blood.

In addition to these three, if you need to have interactive responses for test automations, you might consider Expect, which is based on Tcl. More information about Expect can be found in Ref [18].

Documentation Generator

Every time when I open Matlab (http://www.mathworks.com) for number crunching, I am fascinated by its Publish feature. If you write the comment in a certain format for the Matlab .m file and click the Publish button on the editor toolbar, Matlab will automatically generate a nicely formatted HTML document based on those comments.

In general, the Publish feature you see in Matlab is called a *documentation generator*. It basically means you can write the comment in some mark-up language[2], and a tool (a documentation generator) will extract and render those mark-up languages into nicely formatted documents.

The idea of a documentation generator is not confined to script languages. However, I've found that script languages are often the loyal customers of the document generator. That's the main reason I decided to discuss this topic here.

One thing you can do to generate documents automatically is use a generic documentation generator, such as Doxygen (Ref [6]). Doxygen supports a wide variety of languages, such as C, C++, Tcl, etc. (Amazingly, Doxygen even supports VHDL.) To use Doxygen, the language's comment has to follow specific annotation format, and Doxygen will generate a HTML document based on those comment annotations.

With that being said, script languages often have their own flavor of documentation systems. And more or less, the same can be said for those script language mentioned previously:

- *TclDoc*: I didn't find any built-in features for Tcl that could generate documents based on formatted comments. And the closest thing I can find is a tool called TclDoc (Ref [4]). It offers similar functions for Tcl as Matlab publish does for an *m* file.

- *POD for Perl*: Perl supports a markup language called POD (Plain Old Documentation) in its comments. Details about POD can be found in Ref [5]. Users can use the `perldoc` command to extract and render the POD into nicely formatted text.

- *Python DocStrings and reStructuredText*: Python does go extra miles in this regard. First of all, Python supports a documentation convention called DocStrings (Ref [7]), which starts and ends with triple quotations. Users can use the Help command under Python interactive mode to extract those DocStrings and generate a nicely formatted document.

  However, DocStrings is not just a way of commenting. If you're working with the doctest module (Ref [8]), DocStrings can be used for BIST (Built-In Self-Test) by embedding interactive examples in DocStrings. And among those script languages aforementioned, the integration of documentation and verification is a very unique feature for Python.

- In addition, Python's technical documents (User Guide, Reference Manual, etc.) are written in a markup language called reStructureText (a fruit of the DocUtils project (Ref [11])). Kind like HTML, reStructureText (Ref [10], with .rst extension) by itself is just a text file with lots of markup tags. In order to present its content in a more reader-friendly format, the reStructureText file usually needs to be rendered into other popular formats, such as HTML or PDF. And Sphinx (Ref [9]) is one of the tools that can do the job. (On Windows, Sphinx can be run under Cygwin.)

Summary

This chapter discussed C++ language and script languages.

C++ is not merely an enhancement to C. It also offers a lot of new concepts and methodologies, such as RAII (Resource Acquisition Is Initialization), OOP (Object Oriented Programming), etc. However, abuse of its features can lead to bloated code and unexpected behaviors. Personally I always try to stay away from meta-programming and exceptions during embedded system development.

In addition, script languages can be handy tools for automated testing. Thanks to regular expression, script languages are also good at text processing, like analyzing intermediary results and log files.

---

[2]In its simplest form, the markup language can be a certain format convention or some predefined tags.

References

1. *Real-Time C++, Efficient Object-Oriented and Template Microcontroller Programming.* Chris Michael Kormanyos, Springer-Verlag Berlin Heidelberg, 2013
2. *STL Tutorial and Reference Guide, C++ Programming with the Standard Template Library.* David R. Musser, Gillmer J.Derge and Atul Saini, Addison-Wesley, 1996
3. *C++ Templates: The Complete Guide.* David Vandevoorde and Nicolai M. Josuttis, Addison-Wesley, November, 2002
4. TclDoc Info Page (http://th-labs.de/tcldoc/TclDocInfoPage.html)
5. perlpod—the Plain Old Documentation format (http://perldoc.perl.org/perlpod.html)
6. Doxygen (http://www.stack.nl/~dimitri/doxygen/index.html)
7. Docstring conventions (https://www.python.org/dev/peps/pep-0257/)
8. The Python Library Reference, Release 3.6.0a0. Guido van Rossum and the Python development team, Python Software Foundation
9. SPHINX Python Documentation Generator (http://sphinx-doc.org/)
10. reStructuredText Primer (https://docs.python.org/3.1/documenting/rest.html)
11. Docutils: Documentation Utilities (http://docutils.sourceforge.net/)
12. *The C++ Programming Language, 4th Edition.* Bjarne Stroustrup, Addison-Wesley, 2013
13. *Effective Modern C++, 42 Specific Ways to Improve Your Use of C++11 and C++14.* Scott Meyers, O'Reilly Media, 2015
14. C++11: Create 0 to N constexpr array in C++, Stackoverflow http://stackoverflow.com/questions/19019252/c11-create-0-to-n-constexpr-array-in-c
15. Generating Lookup Table Constant Expressions in C++11, Joshua Napoli's blog, The Critical Section, Agile software development and C++ syntactic delights, http://joshuanapoli.com/blog/2012/12/generating-lookup-table-constant-expressions-in-c11/
16. *Programming: Principles and Practice Using C++ (2nd Edition).* Bjarne Stroustrup, Addison-Wesley, May, 2014
17. *Practical Programming in Tcl and Tk (4th Edition).* Brent B. Welch, Ken Jones with Jeffrey Hobbs, Prentice Hall PTR, 2003
18. *Exploring Expect.* Don Libes, O'Reilly Media, December, 1994
19. GITHUB: Construct a C++11 constant expression array table of generated values. Joshua Napoli, https://github.com/joshuanapoli/table
20. *Object-Oriented Programming with ANSI-C.* Axel-Tobias Schreiner, December, 1994
21. "Portable Inheritance and Polymorphism in C." Miro Samek, *Embedded Systems Programming,* December, 1997

CHAPTER 8

Building and Deployment

As a software project approaches release, its mass increases.

—Laws of Software Relativity

When the scale of your system grows, two things will likely happen:

1. In order to deploy your firmware promptly and consistently, you need a formal build process.

2. Multitasking becomes crucial, which may very likely call for an embedded OS.

This chapter offers some useful clues to answer those challenges.

The Build Process

Source code has to be built into binaries before all the magic happens. And it is imperative to have a formal build process in place for every firmware project. Such a process is supposed to track dependency among source files and produce various flavors of images when configurations are given. Most likely, it will boil down to a makefile, which is the centerpiece of this section. However, due to the plethora of build tools, the makefile is not the only option on the table. Other options will also be explored, and the pros and cons of them will be discussed in detail.

Makefile Basics

The makefile is interpreted by the Make utility, which has a history going back to as early as the 1970s. Given its longevity, the Make utility has produced a handful of derivative versions over the years. For all practical purposes, the rest of the book will discuss makefiles for GNU Make only and present a few useful templates for projects of small and medium scale.

The Make utility has a long list of syntax rules, many of which can be convoluted and tedious. It will be too pandemic to discuss all of them here. For those of you who are ambitious enough to become the master of the Make utility, Ref [3] is a good place to start. At a minimum, you should be familiar with the issues in the following sections.

© Changyi Gu 2016
C. Gu, *Building Embedded Systems*, DOI 10.1007/978-1-4842-1919-5_8

Targets and Prerequisites

The backbone of the makefile is the dependency rule, which is illustrated in Listing 8-1. Both the targets and the prerequisites are files on the disk, and wildcard can also be used to match filenames. (Phony target is an exception, which will be discussed later.) Basically, it means the targets are dependent on those prerequisites. If any of the prerequisites is updated[1], Make utility will try to update the target as well by executing the corresponding shell commands.

The shell commands that are listed under the dependency should follow a certain format:

- There should be a leading Tab character in front of each shell command. Replacing the tab with blank space will produce errors, although tab and spaces look the same in most text editors.

- Each shell command that is invoked by the Make utility will be echoed onscreen. You can put an @ sign in front of that command to disable the auto-echo.

- The Make utility will quit if the shell command returns a non-zero value. You can put a - sign in front of that command to let the Make utility ignore the error and move on[2].

- You can use $@ and $< to refer to targets and prerequisites, respectively.

Listing 8-1. Targets and Prerequisites

```
targets : prerequisites
    shell commands ...
```

Example

```
%.o : %.c
    -@echo "===> Building $@"
    @$(CC) $(CFLAGS) -c -o $@ $<
```

Variable Expansion

You can use variables in your makefiles. Depending on the way variables are assigned, there are two kinds of them: *simply-expanded* and *recursively-expanded*. Simply-expanded variables get to determine their values immediately when they are being assigned, while recursively-expanded ones have to wait until the moment when they are being referenced. In order to distinguish one from the other, the two kinds of variables use := (for simply-expanded) and = (for recursively-expanded) for assignment operation. As demonstrated in Listing 8-2, $(TOPDIR) is a simply-expanded variable while $(dependency) is a recursively-expanded one. As you can see, $(obj) might change its value before $(dependency) is being used. So it makes more sense to defer to $(dependency)'s expansion.

---

[1]The Make utility will check the timestamps to determine whether targets are older than any of the prerequisites.
[2]Another way to process the return value is to use the $$? variable, which will be demonstrated in later sections.

Listing 8-2. Variable Expansion

```
TOPDIR := $(shell pwd)

export TOPDIR

dependency = $(patsubst %.o,%.d,$(obj))
...

ifneq "$(MAKECMDGOALS)" "clean"
   -include $(dependency)
endif
```

Phony Target

As mentioned early, a target is supposed to match the corresponding file on the disk. However, this is not always the case. With the .PHONY directive, you can stop such a filename matching by declaring the target as a phony target. The Make utility will always treat a phony target as out-of-date. As demonstrated in Listing 8-3, both all and clean are phony targets. make all will trigger the processing of all object files before link is conducted, while make clean will prepare for a fresh new build by removing files generated in the previous iteration.

Listing 8-3. Phony Target

```
obj = test/test_main.o \
      common/CRC32.o \
      common/about.o \
      align/align.o

target = demo.exe

...

all: $(obj)
   @echo "====> Linking $(target)"
   @$(LD) $(LDFLAGS) $(obj) -o $(target)
   @chmod 755 $(target)

clean :
   -@rm -vf $(target)
   -@find . -type f \( -name "*.d" -o -name "*.o" \) -exec rm -vf {} \;

.PHONY: clean all
```

Simple Makefile

From this section forward, I will try to provide a few makefile templates for practical use. I will start small by organizing some files mentioned in early chapters into the source tree structure shown in Figure 8-1. Within the top folder, there are four directories. All the common header files, such as debug.h, are stored in the common/include folder, and they will be included in the whole project.

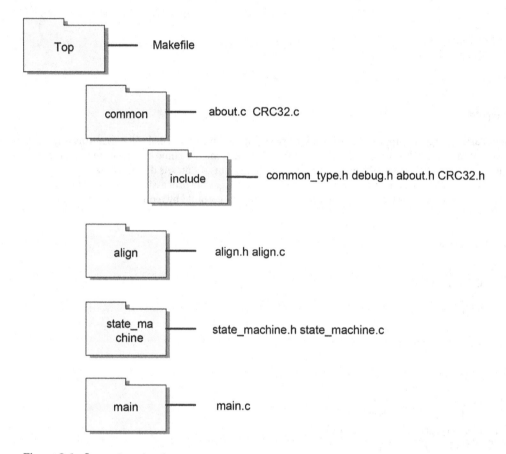

Figure 8-1. *Source tree structure*

Given the source tree structure, you can have a simple makefile as illustrated in Listing 8-4:

- Specify cross-compiler, cross-assembler, and cross-linker in $(CC), $(AS), and $(LD) variables, respectively.

- Specify all the compiler options in $(CFLAGS). These options should include header file search paths and optimization flags. All the global macro definitions, such as VMAJOR, VMINOR, and DEBUG, are supposed to be placed here as well.

- By the same token, you specify all necessary link options in $(LDFLAGS).

- Specify all the modules (object names) in $(obj) variable.

- Specify the target image name in $(target) variable.

- The phony target all will kick-start the compiling for all objects, which will in turn trigger the %.o : %.c rule.

- With the help of a wildcard, the %.o : %.c rule will trigger the compiling for those modules listed in $(obj) variable.

- The phony target clean will kick-start shell scripts to remove files generated in the previous building cycle.

Listing 8-4. Simple Makefile

```
CC = gcc
AS =
LD = gcc

CFLAGS += -I. -Icommon/include -Istate_machine -Ialign -g -W -DVMAJOR=1 -DVMINOR=3g
-DDEBUG -DCONSOLE_PRINT=printf -DABOUT_CRC32=0x68DAC604

LDFLAGS +=

#== Put all the object files here
obj = main/main.o \
      common/CRC32.o \
      common/about.o \
      align/align.o \
      state_machine/state_machine.o

target = demo.exe

all: $(obj)
   @echo "====> Linking $(target)"
   @$(LD) $(LDFLAGS) $(obj) -o $(target)
   @chmod 755 $(target)

%.o : %.c
   @echo "===> Building $@"
   @$(CC) $(CFLAGS) -c -o $@ $<

clean :
   -@rm -vf $(target)
   -@find . -type f \( -name "*.d" -o -name "*.o" \) -exec rm -vf {} \;

.PHONY: clean all
```

The simple makefile in Listing 8-4 serves as a good starting point for small projects, except for one thing: It does not handle dependency among source files. For example, debug.h is included by many sub-modules, such as align.c, about.c, and main.c. However, modifying debug.h does not automatically trigger the rebuilding of these .c files, because such a dependency is not reflected in Listing 8-4. Since source files are in a constant state of changing, and header files could include other headers files in a nested fashion, it can be quite labor-intensive to extract dependencies manually. Fortunately, this dilemma can be solved with the help of compiler option -M.

Makefiles that Handle Dependency

For each source file, C compilers like gcc can produce a list of include files[3] by using the -M option, as demonstrated in Listing 8-5. This list is dumped into a corresponding .d file where the *d* stands for dependency. The sed command in Listing 8-5 is there to correct the output list with directory information, whose purpose will be explained soon. Given all that, each .c file can generate a corresponding .d file that contains the full dependency information.

Listing 8-5. main.c and List of Include Files

```
// main.c
...

#include "stdio.h"
#include "about.h"
#include "align.h"

int main (void)
{
    about();
    alignment_test();

    return 0;
} // End of main()
```

```
$ gcc -I. -Icommon/include -Istate_machine -Ialign -g -W -DVMAJOR=1 -DVMINOR=3g
-DDEBUG -DCONSOLE_PRINT=printf -DABOUT_CRC32=0x68DAC604 -M main/main.c | sed
"s,main.o\s*:,main/main.o :," > main/main.d

$ cat main/main.d

main/main.o : main/main.c /usr/include/stdio.h /usr/include/_ansi.h \
 /usr/include/newlib.h /usr/include/sys/config.h \
 /usr/include/machine/ieeefp.h /usr/include/sys/features.h \
 /usr/include/cygwin/config.h \
 /usr/lib/gcc/i686-pc-cygwin/4.5.3/include/stddef.h \
 /usr/lib/gcc/i686-pc-cygwin/4.5.3/include/stdarg.h \
 /usr/include/sys/reent.h /usr/include/_ansi.h /usr/include/sys/_types.h \
 /usr/include/machine/_types.h /usr/include/machine/_default_types.h \
 /usr/include/sys/lock.h /usr/include/sys/types.h \
 /usr/include/machine/types.h /usr/include/cygwin/types.h \
 /usr/include/sys/sysmacros.h \
 /usr/lib/gcc/i686-pc-cygwin/4.5.3/include/stdint.h /usr/include/stdint.h \
 /usr/include/endian.h /usr/include/bits/endian.h /usr/include/byteswap.h \
 /usr/include/sys/stdio.h /usr/include/sys/cdefs.h common/include/about.h \
 common/include/common_type.h common/include/debug.h align/align.h \
 common/include/common_type.h common/include/debug.h \
 state_machine/state_machine.h
```

---

[3]Nested includes are also covered in this list.

By including these .d files, makefile will be able to determine the exact scope of source files to rebuild after a header file is modified. Thus, the simple makefile in Listing 8-4 can be transformed into the one in Listing 8-6 to support dependency.

Listing 8-6. Makefile that Handles Dependency

```
CC = gcc
AS =
LD = gcc

CFLAGS += -I. -Icommon/include -Istate_machine -Ialign -g -W -DVMAJOR=1 -DVMINOR=3g
-DDEBUG -DCONSOLE_PRINT=printf -DABOUT_CRC32=0x68DAC604
LDFLAGS +=

#== Put all the object files here
obj = main/main.o \
      common/CRC32.o \
      common/about.o \
      align/align.o \
      state_machine/state_machine.o

target = demo.exe

all: $(obj)
    @echo "====> Linking $(target)"
    @$(LD) $(LDFLAGS) $(obj) -o $(target)
    @chmod 755 $(target)

%.o : %.c
    @echo "===> Building $@"
    @echo "============> Building Dependency"
    @$(CC) $(CFLAGS) -M $< | sed "s,$(@F)\s*:,$@ :," > $*.d
    @echo "============> Generating OBJ"
    @$(CC) $(CFLAGS) -c -o $@ $<; \
    if [ $$? -ge 1 ] ; then \
            exit 1; \
    fi
    @echo "-----------------------------------------------------------------------"

dependency = $(patsubst %.o,%.d,$(obj))

ifneq "$(MAKECMDGOALS)" "clean"
    -include $(dependency)
endif

clean :
    -@rm -vf $(target)
    -@find . -type f \( -name "*.OBJ" -o -name "*.d" -o -name "*.o" -o -name "*.LST" -o -name
"*.M51" \
   -o -name "*.hex" -o -name "*.rsp" \) -exec rm -vf {} \;

.PHONY: clean all
```

As you can see, the enhancements made in Listing 8-6 are as follows:

- Dependency information will be extracted from the .c file and dumped into a .d file before the object is generated.

- For each .c file, its corresponding .d file will be included by the makefile. The default dependency information produced by the compiler does not include a directory path. That's what sed "s,$(@F)\s*:,$@ :," is there for—to correct this behavior, since a directory path is required when .d files are being included.

- When the object is generated, the makefile will determine the next step of action based on the compiler's return value. By default, a compiler should return a non-zero value only when there is compiling error. However, some compilers, like Keil C51 (Ref [5]), will return 1 on warning and 2 on error. To let the makefile only stop on compiling errors rather than on warnings, use the automatic variable $$? to convert the compiler's return values.

For all practical purposes, Listing 8-6 can be a good makefile template for small projects. Here my definition of "small" is a project whose total number of source files is less than 200 and total number of sub-folders is less than 20. If your project grows far beyond that point, you might consider deploying recursive makefiles.

Recursive Makefiles

Rome wasn't built in a day, and neither is any meaningful embedded system. The only viable approach to building anything relatively big is divide-and-conquer. You start off with those basic building blocks and then test them and make sure they are solid. On top of those basic blocks, you can create bigger modules. You test them so they can be used as solid building blocks as well. And you repeat those aforementioned steps recursively until you reach the top. As far as development is concerned, it implies the following:

- The project is composed of multiple sub-modules.

- There could be dependency among those sub-modules, and such a dependency should be reflected in build process as well. Basically it means that some sub-modules have to be built before others in order not to violate the dependency rule. Note that mere symbol reference from one module to another does not necessarily constitute build dependency. The two modules can be built in any chosen order and linked at the last stage. However, build dependency does exist if module A is referencing files from module B and these files are dynamically generated by module B in its build process. Consequently, you can assert that A is dependent on B and should be built after B.

- Since the whole project is developed from bottom up, the building and testing should be carried out at both the sub-module level and the top level.

To accommodate these issues, I suggest the project source tree and makefiles being organized in the following manner:

- Each sub-module should have a corresponding sub-folder in source tree, and it is better to have a flat structure for the source tree. The dependency among sub-modules should be reflected in the makefile instead of the source tree structure.

- Since each sub-module should be built and tested independently, you can put two makefiles in each sub-folder. One is named literally as Makefile, and it is responsible for building the sub-module. The other can be called Makefile_Test, and it is for building the test fixture.

- As you can imagine, most makefiles look the same, which are the close cousins of Listing 8-6. To avoid code duplication, the greatest common factor of them, namely the bulk of Listing 8-6, can be extracted into a file called `Rules.mak.` `Rules.mak` resides in the top level. All the makefiles in the project are supposed to include `Rules.mak`. Generally, makefiles differ from each other only by specifying the distinctive objects and targets involved.

- As stated earlier, dependency among sub-modules has to be accommodated. However, dependency can grow into something that involves multiple layers of assorted modules. Assuming there is no mutual or cyclic dependency[4], you can describe module dependency by using DAG (Directed Acyclic Graph), as illustrated in Figure 8-2. The longest path in Figure 8-2 is the length of four (Top -> A1 -> B0 -> C0 -> D0 and Top -> A1 -> B0 -> C1-> D0). Based on each module's longest distance from top, the DAG in Figure 8-2 can be divided into four layers. For a medium-scale project, the longest path in its dependency DAG is usually no more than four. So I will design my template of recursive makefiles based on this assumption and name the four layers of sub-modules `submodules`, `submodules_sofa`, `submodules_carpet`, and `submodules_floor`. (As you can see, submodules sit on the sofa, and the sofa is on top of the carpet with the floor underneath. Hopefully I don't have to add one more layer for the basement.) These names will correspond to respective variables in the makefile to implement sub-module dependency.

---

[4]Mutual or cyclic dependency could happen under complex circumstances. For example, both module A and module B could reference each other's configuration header files `config_A.h` and `config_B.h`, while these configuration header files are supposed to be generated dynamically during the build process. Due to their complexity, mutual dependency and cyclic dependency are usually solved on an ad hoc basis, and they will not be discussed in this book.

157

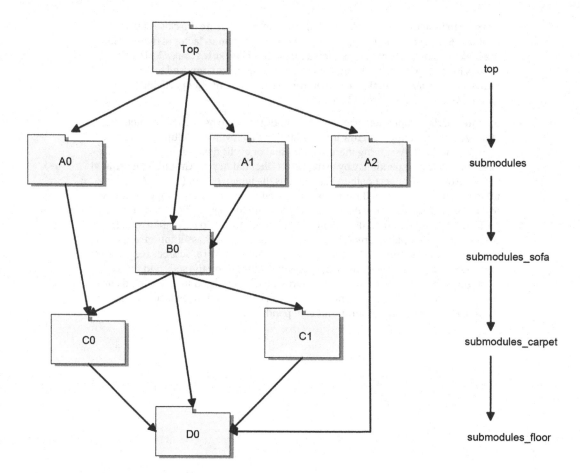

Figure 8-2. *Describe dependency with DAG*

Now let me use the source tree structure in Figure 8-1 to show you how recursive makefiles can be employed to tackle the same problem. To use recursive Make, there should be makefiles in each sub-folder. As stated previously, a test fixture can also reside in the sub-folder to carry out a unit test. So the source tree in Figure 8-1 will evolve into the one in Figure 8-3 with more files in it.

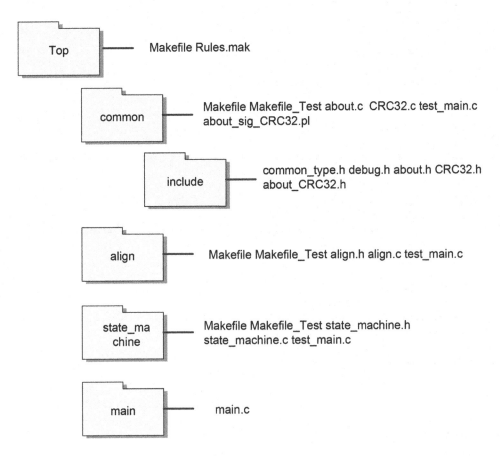

Top — Makefile Rules.mak

common — Makefile Makefile_Test about.c CRC32.c test_main.c
about_sig_CRC32.pl

include — common_type.h debug.h about.h CRC32.h
about_CRC32.h

align — Makefile Makefile_Test align.h align.c test_main.c

state_ma chine — Makefile Makefile_Test state_machine.h
state_machine.c test_main.c

main — main.c

Figure 8-3. *Source tree for recursive make*

Compare Figure 8-3 to Figure 8-1 and you will find that the source tree folder structure is the same. The only difference is that each sub-folder now has three more files: Makefile, Makefile_Test, and test_main.c. Makefile is for building the sub-module, while Makefile_Test and test_main.c are part of the test fixture. Rules.mak resides in the top folder, and it is supposed to be included by all the makefiles.

Listing 8-7. Rules.mak

```
CFLAGS += -I. -I$(TOPDIR)/common/include
CFLAGS += -g -W -DVMAJOR=1 -DVMINOR=3g -DDEBUG -DCONSOLE_PRINT=printf

all: $(submodules) $(obj)
    @if [ -n "$(target)" ] ; then \
        echo "====> Linking $(target)";\
        $(CROSS_LD) $(LDFLAGS) $(obj) $(link_obj) -o $(target); \
    fi

    @if [ -n "$(post_link_action)" ] ; then \
        echo "====> $(post_link_action)"; \
    fi
```

```
    @$(post_link_action)

$(submodules) : $(submodules_sofa)
    @$(MAKE) -C $@
    @echo "==========================================================================="
    @echo

$(submodules_sofa) : $(submodules_carpet)
    @$(MAKE) -C $@
    @echo "==========================================================================="
    @echo

$(submodules_carpet) : $(submodules_floor)
    @$(MAKE) -C $@
    @echo "==========================================================================="
    @echo

%.o : %.c
    @echo "===> Building $@"
    @echo "============> Building Dependency"
    @$(CROSS_CC) $(CFLAGS) -M $< | sed "s,$(@F)\s*:,$@ :," > $*.d
    @echo "============> Generating OBJ"
    @$(CROSS_CC) $(CFLAGS) -c -o $@ $<; \
    if [ $$? -ge 1 ] ; then \
        exit 1; \
    fi
    @echo "---------------------------------------------------------------------"

dependency = $(patsubst %.o,%.d,$(obj))

ifeq (,$(filter "clean" "clean_recursive", "$(MAKECMDGOALS)"))
    -include $(dependency)
endif

clean_recursive :
    -@for dir in $(submodules_floor) $(submodules_carpet) $(submodules_sofa)
$(submodules); do \
        $(MAKE) -s -C $$dir clean; \
    done

    -@rm -vf $(target)
    -@rm -vf $(obj)
    -@rm -vf $(patsubst %.o,%.d,$(obj))

clean :
    -@rm -vf $(target)
    -@find . -type f \( -name "*.OBJ" -o -name "*.d" -o -name "*.o" -o -name "*.LST" -o -name
"*.M51" \
                        -o -name "*.hex" -o -name "*.rsp" \) -exec rm -vf {} \;

.PHONY: clean all clean_recursive $(submodules) $(submodules_sofa) $(submodules_carpet)
$(submodules_floor)
```

Listing 8-7 has the bulk of Rules.mak, and comparing Listing 8-7 to Listing 8-6 will reveal the following:

- $(submodules), $(submodules_sofa), $(submodules_carpet), and $(submodules_floor) are added to Rules.mak to enforce the building order. These variables are also listed as phony targets. For each dependency rule of these variables, $(MAKE) -C is used to iterate all the sub-folders involved.

- To make Rules.mak more generic, $(post_link_action) is added to accommodate script processing after the link is done.

- $(obj) and $(target) are not specified in Rules.mak. Instead, they are supposed to be set up by the makefile.

The top/common/include/about_CRC32.h file in Figure 8-3 contains the CRC32 checksum of the about signature, and it is produced dynamically by a Perl script (the script about_sig_CRC32.pl will derive the about signature from the binary files and calculate the CRC32). Other modules, such as state_machine and main, will include this header file. Thus there is dependency and the common module should be built before the others. The whole DAG for Figure 8-3 is shown in Figure 8-4.

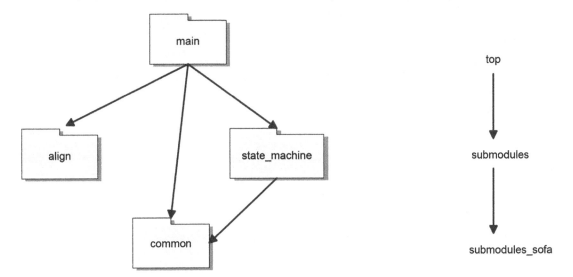

Figure 8-4. *Example DAG*

As you can deduce from Figure 8-4, $(submodules) and $(submodules_sofa) will be employed in the makefile to force the build order among modules. Thanks to Rules.mak, the top-level makefile can be more succinct, as shown in Listing 8-8.

Listing 8-8. Top-Level Makefile

```
TOPDIR := $(shell pwd)

export TOPDIR

obj = main/main.o
target = demo
```

```
link_obj = state_machine/state_machine.o common/common.o align/align.o

submodules = align state_machine
submodules_sofa = common

CROSS_LD = gcc

CFLAGS += -I$(TOPDIR)/align -I$(TOPDIR)/state_machine

include $(TOPDIR)/Rules.mak
```

As stated earlier, building and testing should also be carried out at sub-module level. Accordingly, there are two makefiles in each sub-folder: Makefile and Makefile_Test. The former is for building only while the latter is for unit testing. Take sub-module state_machine for example; its Makefile and Makefile_Test files are demonstrated in Listings 8-9 and 8-10, respectively. As you can see, state_machine's dependency on the common module is reflected in Makefile_Test, but not in Makefile. This is because the unit test needs to link all the relevant sub-modules to produce a complete binary. For the recursive make, the top-level makefile will handle the module dependency as a whole.

Listing 8-9. Makefile of state_machine Module

```
obj += state_machine.o
target =

ifndef TOPDIR
    TOPDIR := ..
endif

include $(TOPDIR)/Rules.mak
```

Listing 8-10. Makefile_Test of state_machine Module

```
ifndef TOPDIR
    TOPDIR := ..
endif

obj = test_main.o
target = test_main

link_obj = state_machine.o $(TOPDIR)/common/common.o

submodules += $(TOPDIR)/common $(TOPDIR)/state_machine

CROSS_LD = gcc

include $(TOPDIR)/Rules.mak
```

The recursive makefile architecture demonstrated in this section can serve as a template for medium-scale firmware projects with a reasonable build time. However, for the sake of completeness, be aware that there is also an alternative approach that employs a non-recursive, centralized makefile, as suggested in Ref [4]. The non-recursive makefile will include module-specific information (a file called module.mk) from each sub-folder, and it will have multiple targets to build sub-modules. Personally I am not very enthusiastic about this non-recursive approach, for the following three reasons:

- The problems described in Ref [4] do not apply to the makefile template presented in this section.

- The architecture of a firmware project is NOT always in a static shape throughout its lifecycle. It might be premature to have a centralized makefile when all the possible avenues are still being explored at the planning stage. It runs counter to the ideas of divide-conquer and bottom-up approach that I mentioned earlier.

- Oftentimes, your project will utilize modules provided by a third party. These modules are usually self-contained, with their own makefiles out of the box. It would be reinventing the wheel to impose non-recursive makefiles on those modules.

Thus from the perspective of practicality, I don't recommend such a non-recursive solution to readers. Instead, consider the makefile template presented in this section for your new project.

Makefile with Kernel Configuration Language

One thing not included in the previous template is the GUI-based configuration. The kernel configuration language that accompanies the Linux kernel offers a nice solution in this regard.

Configuration Process

To survive and beat your competitors, your project team has to answer calls from a wide variety of customers, in a prompt fashion. More often than not, those demands from customers are only slightly different from each other. For example, most OEM customers want to customize the display of product names and versions. Some customers, in order to differentiate themselves from the pack, may choose to include certain modules (with cool features supposedly) offered by a third-party. Consequently, the build process has to be flexible enough to accommodate those demands. And under such fast-spinning circumstances, the surviving guide usually suggests building the firmware from the same baseline, but with macro definitions to conduct conditional compiling. The build process is basically divided into two stages: configuration and compile/link. The configuration stage will produce a makefile, which is fed into the second stage.

One way to do this configuration is, of course, to manually modify the makefile every time when a new customer demand comes in. However, when your business is booming and the list of customers grows, the manual configuration job quickly becomes tedious and error-prone. Thus most companies will develop some scripts to automate the configuration process. The drawback of in-house scripts is obvious: These scripts are custom-made and offer no generic value to other projects.

Fortunately, since configuration is such a pandemic issue, there are generic solutions. One of them is to use the kernel configuration language, which is discussed in detail in this section.

The Linux Kernel Configuration Language

The Linux kernel finds itself with hundreds of sub-modules, and it is targeted to be run on more than 20 different CPU architectures. Such a huge variety makes configuration an acute issue. A special kernel configuration language was thus developed, and it has undergone some major revamps after kernel 2.5 (Ref [6]). The new kernel configuration language provides a generic solution that is not confined only to the Linux kernel. And for the rest of the book, the new kernel configuration system will be referred to as *Kconfig*. Kconfig is an independent part by itself, and any project that needs a configuration process can borrow a page from Linux kernel by using Kconfig.

163

As shown in Figure 8-5, the configuration process needs to provide config options (manifested in the kernel configuration language) to Kconfig, and these config options are usually saved in a text file named Config.in or Kconfig. Based on this text file, Kconfig will present four different configuration interfaces: command line, text-mode menu, GTK+ based GUI, and Qt based GUI. These four configuration interfaces correspond to the four makefile targets respectively: config, menuconfig, gconfig, and xconfig.

No matter what interface you choose, all of them will save your configuration in a file named .config. This .config file can be included in your makefile or Rules.mak, which will then produce conditional compiling macros for code building (see Figure 8-5).

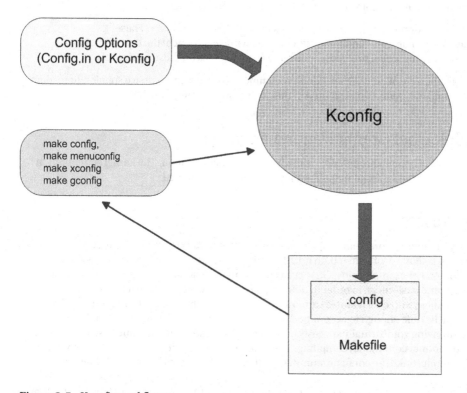

Figure 8-5. *Kconfig workflow*

The support of GUI makes Kconfig stand out and shine. I think it should be the first thing to consider if your firmware project merits some sort of configuration process.

Integrate Kconfig into Firmware Project

To demonstrate Kconfig integration, I will recycle the example shown earlier by adding a configuration process to it. Assume there is a customer from Mars who put forward the following demands:

- The product name and product version (VMajor and VMinor) should be configurable.

- Since the firmware code might be ported to various CPU architectures, the cross-compiler should also be configurable.

- As shown earlier, `alignment` and `state_machine` are the two major modules in the project. Each module should be individually included or excluded by configuration options.

- There should be a separate option to enable/disable about-signature verification.

- Due to the `common` module's popularity, there should be a dedicated checkbox option to rebuild it.

These demands can thus be manifested by the kernel configuration language (saved as `Config.in`) shown in Listing 8-11. As you can see, it is composed of three menus: About, Tools Setup, and Modules Configuration:

- There are three string entries under the About menu. These entries are used to specify product name, major version, and minor version (for demand #1).

- There are two string entries under the Tool Setup menu to satisfy demand #2. One entry is used to set up the cross-compiler for firmware. The other one is in case a different compiler is used. (Some applications may prefer μClibc over the standard C library. This option is mainly reserved for this purpose.)

- Under the Modules Configuration menu, the `ENABLE_ALIGNMENT_MODULE` and `ENABLE_STATE_MACHINE_MODULE` entries are both dependent on `BUILD_WHOLE_IMAGE`. These are Boolean options and they are designed to meet demand #3.

- Demand # 4 and #5 are answered by Boolean entries `CHECK_ABOUT_SIGNATURE` and `REBUILD_COMMON`, respectively.

- What demonstrated in this example is only a subset of the kernel configuration language, and more detailed information on Kconfig can be found in Ref [7].

Listing 8-11. Config.in

```
mainmenu "Firmware Configuration"

config HAVE_DOT_CONFIG
    bool
    default y


#=========================================================================
#  Firmware Version
#=========================================================================

menu "About"

config PRODUCT_NAME
    string "Product Name"
    default "A Super Product"
    help
          Name of the Product

config VMAJOR_STR
    string "Major Version"
    default "xx"
    help
      Major Version for the firmware
```

```
config VMINOR_STR
    string "Minor Version"
    default "xx"
    help
      Minor Version for the firmware

endmenu

#========================================================================
#  Tools Setup
#========================================================================

menu "Tools Setup"

config CROSS_COMPILER_PREFIX
    string "Cross Compiler prefix"
    default "arm-elf-"
    help
      Cross compiler pre-fix, such as arm-elf-, xscale_le-

config APP_COMPILER_PREFIX
    string "Cross Compiler prefix for App Build"
    default "arm-uclibc-"
    help
      Cross compiler pre-fix, such as arm-uclibc-, for application build.
      This is often used in the case where uclibc is preferred over standard
      C library.

endmenu

#========================================================================
#  Config all the modules in the firmware
#========================================================================

menu "Modules Configuration"

config BUILD_WHOLE_IMAGE
    bool "Build Whole Image"
    default y
    help
          Build Whole Image

config ENABLE_EVERYTHING
    bool "Enable All Modules"
    default n
    depends on BUILD_WHOLE_IMAGE
    select ENABLE_ALIGNMENT_MODULE
    select ENABLE_STATE_MACHINE_MODULE
    help
          To enable all modules
```

```
config ENABLE_ALIGNMENT_MODULE
    bool "Enable Alignment Module"
    default n
    depends on BUILD_WHOLE_IMAGE
    help
            To Enable Alignment Module

config ENABLE_STATE_MACHINE_MODULE
    bool "Enable State Machine Module"
    default y
    depends on BUILD_WHOLE_IMAGE
    help
            To Enable State Machine Module

comment "--------------------------------"
    depends on BUILD_WHOLE_IMAGE

config CHECK_ABOUT_SIGNATURE
    bool "Check About Signature"
    default y
    depends on BUILD_WHOLE_IMAGE
    help
            To check the integrity of about signature at startup

comment "===================================="

config REBUILD_COMMON
    bool "Rebuild Common Module"
    default n
    help
            Rebuild Common Module

endmenu
```

In order to process Config.in, the source tree in Figure 8-3 has to be expanded to include the Kconfig module. As demonstrated in Figure 8-6, the new source tree will have a config folder, which homes the Kconfig module extracted from the Linux kernel. The Config.in file from Listing 8-11 resides at the top level.

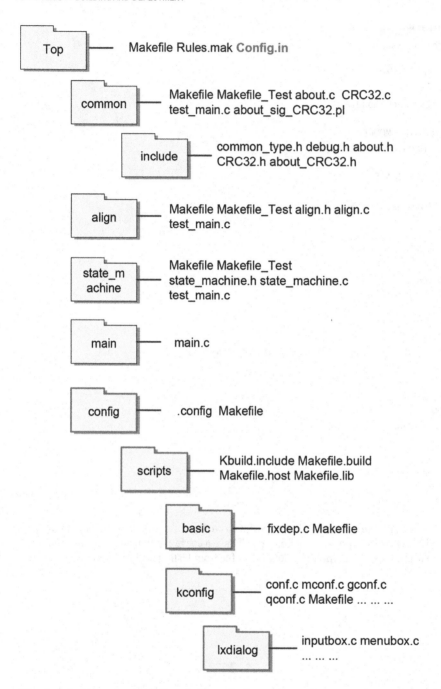

Figure 8-6. *Source tree with Kconfig integrated*

As mentioned earlier, Kconfig supports four different kinds of interfaces that are mapped to respective make targets. Listing 8-11 can thus be tendered in four different ways, as shown in Figures 8-7 through 8-10. The exciting news is that this petty example project now can be configured by a variety of GUIs, which lends it a professional touch.

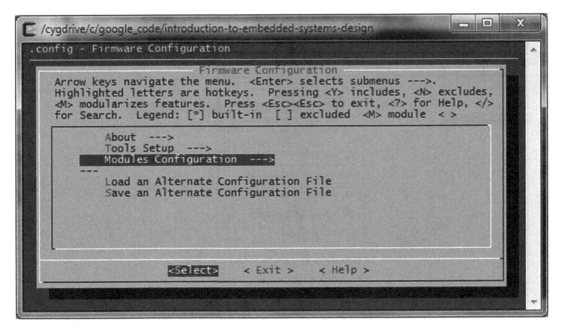

```
/cygdrive/c/google_code/introduction-to-embedded-systems-design                    _  □  X

$ make config
make[1]: Entering directory `/cygdrive/c/google_code/introduction-to-embedded-sy
stems-design/config'
make[2]: Entering directory `/cygdrive/c/google_code/introduction-to-embedded-sy
stems-design/config'
  gcc -Wp,-MD,scripts/basic/.fixdep.d        -o scripts/basic/fixdep scripts/basic
/fixdep.c
make[2]: Leaving directory `/cygdrive/c/google_code/introduction-to-embedded-sys
tems-design/config'
make[2]: Entering directory `/cygdrive/c/google_code/introduction-to-embedded-sy
stems-design/config'
  gcc  -o scripts/kconfig/conf scripts/kconfig/conf.o scripts/kconfig/zconf.tab.
o
scripts/kconfig/conf --oldaskconfig /cygdrive/c/google_code/introduction-to-embe
dded-systems-design/Config.in
*
* Firmware Configuration
*
*
* About
*
Product Name (PRODUCT_NAME) [A Super Product]
Major Version (VMAJOR_STR) [1] 2
Minor Version (VMINOR_STR) [3g]
```

Figure 8-7. *The make config (command line version)*

```
/cygdrive/c/google_code/introduction-to-embedded-systems-design                    _  □  X

.config - Firmware Configuration
                          ┌──────────── Firmware Configuration ────────────┐
     Arrow keys navigate the menu.  <Enter> selects submenus --->.
     Highlighted letters are hotkeys.  Pressing <Y> includes, <N> excludes,
     <M> modularizes features.  Press <Esc><Esc> to exit, <?> for Help, </>
     for Search.  Legend: [*] built-in  [ ] excluded  <M> module  < >
     ┌────────────────────────────────────────────────────────────────┐
     │       About   --->                                             │
     │       Tools Setup   --->                                       │
     │       Modules Configuration   --->                             │
     │       ---                                                      │
     │       Load an Alternate Configuration File                     │
     │       Save an Alternate Configuration File                     │
     │                                                                │
     └────────────────────────────────────────────────────────────────┘
                  <Select>     < Exit >     < Help >
```

Figure 8-8. *The make menuconfig (text-mode menu version)*

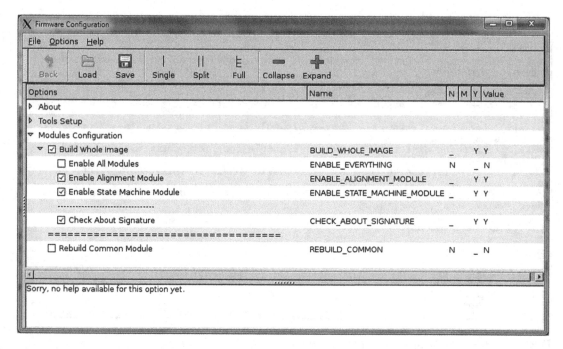

Figure 8-9. *The make gconfig (GTK+ version)*

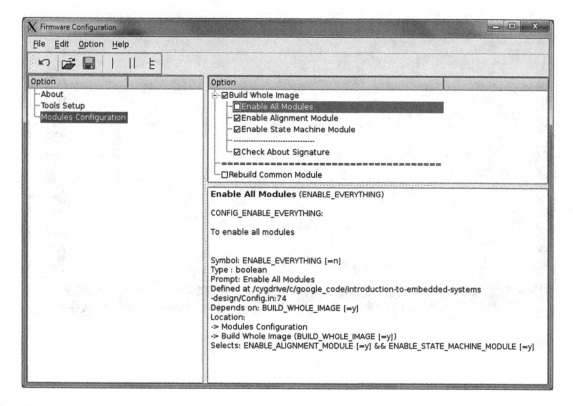

Figure 8-10. *The make xconfig (Qt version)*

No matter what interface you prefer, a file called .config will be saved in the config folder afterward. For this project, the .config may look like something in Listing 8-12.

Listing 8-12. .config

```
#
# Automatically generated file; DO NOT EDIT.
# Firmware Configuration
#
CONFIG_HAVE_DOT_CONFIG=y

#
# About
#
CONFIG_PRODUCT_NAME="A Super Product"
CONFIG_VMAJOR_STR="1"
CONFIG_VMINOR_STR="3g"

#
# Tools Setup
#
CONFIG_CROSS_COMPILER_PREFIX="arm-elf-"
CONFIG_APP_COMPILER_PREFIX="arm-uclibc-"

#
# Modules Configuration
#
CONFIG_BUILD_WHOLE_IMAGE=y
# CONFIG_ENABLE_EVERYTHING is not set
CONFIG_ENABLE_ALIGNMENT_MODULE=y
CONFIG_ENABLE_STATE_MACHINE_MODULE=y

#
# --------------------------------
#
CONFIG_CHECK_ABOUT_SIGNATURE=y

#
# ====================================
#
# CONFIG_REBUILD_COMMON is not set
```

As you can see from Listing 8-12, .config is nothing but a bunch of variable definitions, and they are supposed to be included by Rules.mak or by the top-level makefile. The Rules.mak file in Listing 8-7 is now transformed into Listing 8-13. (Due to the constraint of real estate space, only the part relevant to Kconfig is shown in Listing 8-13.) The new Rules.mak file will include .config at the beginning, and object files should also be dependent on .config, so that they will be rebuilt when new configuration is saved.

Listing 8-13. Rules.mak that Supports Kconfig

```
...

config_targets ?= menuconfig config xconfig gconfig
config_folder_name ?= config

#check to see if we need to include ".config"
ifneq ($(strip $(CONFIG_HAVE_DOT_CONFIG)),y)

    ifeq (,$(filter $(config_targets), "$(MAKECMDGOALS)"))
        -include $(TOPDIR)/$(config_folder_name)/.config
    endif

endif

ifeq ($(strip $(CONFIG_HAVE_DOT_CONFIG)),y)
    config_file := $(TOPDIR)/$(config_folder_name)/.config
else
    config_file :=
endif

...

%.o : %.c $(config_file)
   @echo "===> Building $@"
   @echo "============> Building Dependency"
   @$(CROSS_CC) $(CFLAGS) -M $< | sed "s,$(@F)\s*:,$@ :," > $*.d
   @echo "============> Generating OBJ"
   @$(CROSS_CC) $(CFLAGS) -c -o $@ $<; \
   if [ $$? -ge 1 ] ; then \
        exit 1; \
   fi
   @echo "--------------------------------------------------------------------"

...
```

A top-level makefile should also be revised to invoke Kconfig when .config is missing, as illustrated in Listing 8-14. Compare Listing 8-14 to Listing 8-8; the new makefile is now conspicuously bigger for the following reasons:

- A big ifeq-else forms the skeleton of new makefile, and the else branch is responsible for invoking the Kconfig when .config is not detected.

- The ifeq branch is the mainstay of the makefile. Most variables from .config will be processed in this branch, translated into macro definitions, and passed along to source files as compiler options. However, variables like product name and version will be processed in Rules.mak instead of makefile, because they are applicable across the board.

Listing 8-14. Top-Level Makefile with Kconfig Support

```
TOPDIR := $(shell pwd)
export TOPDIR

config_targets := menuconfig config xconfig gconfig
config_folder_name := config

# Pull in the user's configuration
ifeq (,$(filter $(config_targets),$(MAKECMDGOALS)))
   -include $(TOPDIR)/$(config_folder_name)/.config
endif


##########################################################################
ifeq ($(strip $(CONFIG_HAVE_DOT_CONFIG)),y)

        link_obj += main/main.o
        submodules += main

        ifeq ($(strip $(CONFIG_BUILD_WHOLE_IMAGE)),y)
            target := demo
        endif

    #++++++++++++++++++++++++++++++++++++++++++++++++++++++++++++++++++++++
        ifeq ($(strip $(CONFIG_ENABLE_ALIGNMENT_MODULE)),y)
            submodules_sofa += align
            link_obj += align/align.o
            CFLAGS += -DCONFIG_ENABLE_ALIGNMENT_MODULE
        endif
    #++++++++++++++++++++++++++++++++++++++++++++++++++++++++++++++++++++++
        ifeq ($(strip $(CONFIG_ENABLE_STATE_MACHINE_MODULE)),y)
            submodules_sofa += state_machine
            link_obj += state_machine/state_machine.o
            CFLAGS += -DCONFIG_ENABLE_STATE_MACHINE_MODULE
        endif
    #++++++++++++++++++++++++++++++++++++++++++++++++++++++++++++++++++++++
        ifeq ($(strip $(CONFIG_CHECK_ABOUT_SIGNATURE)),y)
            CFLAGS += -DCONFIG_CHECK_ABOUT_SIGNATURE
        endif
    #++++++++++++++++++++++++++++++++++++++++++++++++++++++++++++++++++++++
        submodules_carpet = common
        link_obj += common/common.o

        CROSS :=
        CROSS_LD := gcc
        export CROSS
    #++++++++++++++++++++++++++++++++++++++++++++++++++++++++++++++++++++++
        ifeq ($(strip $(CONFIG_REBUILD_COMMON)),y)
            submodules_floor = clean_common
            submodules_floor_build_command = \
```

```
                @echo "===> clean common module";\
                make -C common clean;\
                echo "-----------------------------------------------------"

        endif
    #++++++++++++++++++++++++++++++++++++++++++++++++++++++++++++++++++++++++++++
        export CFLAGS
        include $(TOPDIR)/Rules.mak


###############################################################################
else # need to be configured

$(config_targets):
    @make -C $(config_folder_name) $(MAKECMDGOALS)

.PHONY: $(config_targets)

endif
```

To summarize, if you want to integrate Kconfig into your build process, the Rules.mak and makefile demonstrated in this section can serve as a template for your reference. The kernel configuration language is very intuitive, and the support of GUI is its most valuable asset. That's why I strongly recommend it.

Other Ways to Implement a Build Process

Jumping on the bandwagon is not a virtue, nor is basking in the comfort zone. The main reason I've included this section is an attempt to not be single-minded, as the old saying goes: one man's limerick is another man's clue trail. The build process that has been illustrated so far centers on makefile, with Kconfig as its configuration sub-process. Personally, I think this is a handy approach for small- and medium-scale firmware projects. However, as you can imagine, there are many other options that you can explore when it comes to a build process. And those alternative approaches will be introduced in this section.

Among those alternative approaches, my personal favorite is SCons, which I will elaborate with practical details. For other options that are introduced in this section, although they may not be the best match for building firmware, they still have their merits.

Automake and Autoconf

Automake (Ref [8]) and Autoconf (Ref [9]) are part of the GNU build system (Autotool) that comprises Automake, Autoconf, Libtool, and Gnulib. Their main purpose is to make it easy to build and distribute packages across a wide variety of platforms. They can fathom out the differences among various platforms and dynamically produce a makefile to resolve those differences. In addition, chores like dependency checks can also be handled by these dynamically generated makefiles.

As illustrated in Figure 8-11, to use Automake and Autoconf, users have to prepare two files: Makefile.am and configure.ac. (Optionally, Makefile.am can include another file called aclocal.m4, which is generated by the aclocal command.) Automake will produce an output file called Makefile.in based on Makefile.am, while Autoconf will generate a script called configure. Afterward, the configure script will be executed to produce a makefile based on Makefile.in. Differences among various platforms will be reflected in this produced makefile.

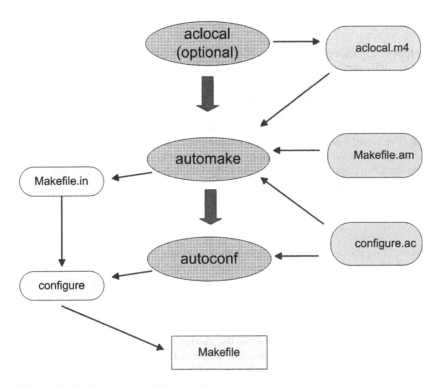

Figure 8-11. *Automake and Autoconf*

Automake and Autoconf can be used to distribute source code packages to a wide variety of hosting systems. The produced makefile will have targets like all and install to facilitate code building and installation. However, being able to build across multiple hosting systems is less of a concern for firmware development, since the chance of moving to a different cross-building environment is quite low over the entire project lifecycle.

With a long list of predefined macros, preparing Makefile.am and configure.ac can be a daunting task. (Part of this complexity comes from the GNU M4 (Ref [10]), which is the macro language processor employed by Autoconf.) That's why I don't think Automake and Autoconf are suitable for firmware projects, unless you want to distribute your source packages to a variety of hosting environments.

CMake

CMake (Ref [11]) is a cross-platform build system that supports configuration and dependency detection. To achieve cross-platform use, CMake works with its own flavor of makefile, called CMakeList.txt. The output of CMake is makefiles or project files that your development tool can recognize, such as a Unix makefile, Visual Studio Project File, Borland makefile, etc. CMake relies on these makefiles or project files to do the final build. In that sense, CMake is similar to Automake/Autoconf.

The prerequisite of have CMakeList.txt means developers have to master a new script language in order to use CMake, which is a drawback to many including me. (I'm not ashamed to admit that I have "language inertia".) And it takes a while to iron out all the quirks before you can write CMakeList.txt with sophistication. Since most firmware projects stick with a single platform (host machine) throughout the whole development cycle, the cross-platform is less attractive to embedded system development.

BJAM

JAM is a build tool first introduced by Perforce Software, Inc. (Ref [12]). And shortly after its inception, several refinements ensued. One of them made its way into the toolset of the Boost Library (Ref [13]), which is now called Boost Build (Ref [14]) or BJAM. Like Autoconf and CMake, BJAM also relies on its own script language (Boost Jam) to conduct the build flow, but it goes further by completely replacing the makefile with a Boost Jam file (Jamroot and Jamfile). In that sense, it is a self-contained build tool.

BJAM supports most mainstream compilers, like GNU C/C++ and Visual C++, and is popular among C++ programmers. Personally, I've found it works well if you work with Boost and C++. However, once you walk out of that setting, you will find yourself a lot of hoops to jump through before you can make a successfully customized build. And like CMake, mastering a new script language only for a build is not favorable to many, and it does not have GUI configuration either.

SCons

As you can see from previous sections, those who want to replace the traditional makefile have to pick up a second language. Autoconf relies on GNU M4, CMake uses CMakeList.txt, and BJAM adopts Boost Jam. SCons (Ref [15]) also has invented its own build script. However, unlike the other build tools, SCons' build script is merely a Python script. In other words, SCons did not invent a new language. Instead, it has invented some new Python functions and classes and put them together to form a new build tool.

The dynamics of Python gives SCons great power. SCons supports dependency detection, external commands, multiple variants, the separation of source and build directories, etc. (Ref [16]). And because its build script is pure Python, those who are already familiar with Python can pick it up easily. Personally, I like the flexibility that SCons can offer. *If your embedded project needs a build process other than a make/makefile, I suggest you consider SCons.*

To elaborate on this, I have provided an example project in the companion material, which can be used as templates for future development. The corresponding source tree is shown in Figure 8-12.

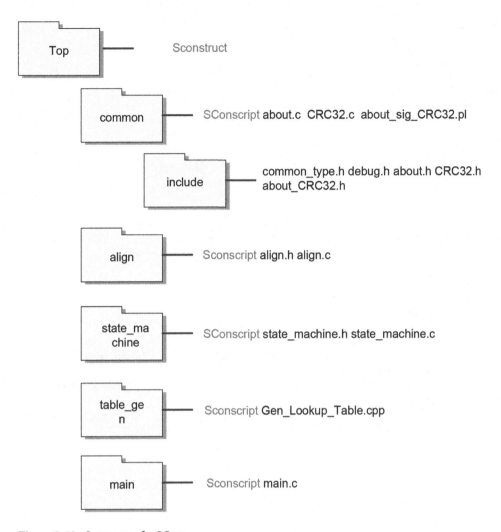

Figure 8-12. *Source tree for SCons*

As you can see, the source tree here bears a striking resemblance to the one shown in Figure 8-3, except for the following:

- At the top folder, the makefile is replaced by a file called SConstruct. At the sub-folder level, makefiles are replaced by files called SConscript. Both SConstruct and SConscript are in fact Python scripts.

- I have added another folder called table_gen on top of Figure 8-3 to show how sub-projects can be customized. And table_gen is based on the example for "lookup table auto-generation" from previous chapters.

For SCons, it will always look for a file called SConstruct at the top level. The main purpose of SConstruct is to set up the environment and build all sub-projects with the respective SConscript. For the example in Figure 8-12, the corresponding SConstruct is shown in Listing 8-15.

Listing 8-15. SConstruct Template

```
...
#=============================================================================
# Compiling Environment set up
#=============================================================================
include      = [".", "#common/include", "#align", "#state_machine"]
c_flags      = []
cpp_flags    = ['-std=c++11']
cc_flags     = ['-O3', '-Wall']
sub_projects = ['state_machine', 'main', 'align', 'common', 'table_gen']
output_dir   = '#output_files'


...
#=============================================================================
# Build Variables set up, use "VERBOSE=yes" for verbose output
#=============================================================================


vars = Variables(None)
vars.AddVariables( ('VMAJOR', 'major version', '1'),
                   ('VMINOR', 'minor version', '3g'),
                   ('PRODUCT_NAME', 'Product Name', 'A SUPER Product'),
                   BoolVariable('VERBOSE',                         'Verbose output', 0),
                   BoolVariable('CONFIG_CHECK_ABOUT_SIGNATURE',    'About Signature
                   enable/disable', 1),
                   BoolVariable('CONFIG_ENABLE_ALIGNMENT_MODULE',  'align module enable/
                   disable', 1),
                   BoolVariable('CONFIG_ENABLE_STATE_MACHINE_MODULE', 'state machine module
                   enable/disable', 1)
                 )
...


#=============================================================================
# setup cross compiler here
#=============================================================================


## default_env.Replace (CC="sdcc")
## default_env.Replace (CXX="arm-g++")
## default_env.Append(ENV = {'PATH' : os.environ['PATH']})
...
```

SConstruct is actually a big file. I've only shown the part that might need to be customized, such as the sub-project folder name, compiler flag, etc. And for this particular example, macro definitions like VMAJOR, VMINOR, and PRODUCT_NAME can be set in the SConstruct directly or provided through the command line.

At the sub-project level, most SConscript files will look like the one shown in Listing 8-16.

Listing 8-16. SConscript Template

```
src_files = ['xxxxxx.c']
obj = Object (src_files)
Return('obj')
```

However, project `table_gen` is an exception. As mentioned in previous chapters, to accommodate large tables, `-ftemplate-depth` needs to be bumped up with a bigger number. The corresponding `SConscript` file looks like Listing 8-17.

Listing 8-17. SConscript with a Customized Build Option

```
src_files = ['Gen_Lookup_Table.cpp']

env = DefaultEnvironment()
obj = Object (src_files, CXXFLAGS = env['CXXFLAGS'] + ['-ftemplate-depth=2048'])

Return('obj')
```

SCons can be installed along with Cygwin, or it can be installed separately, as illustrated in more detail in Ref [16]. For the example at hand, the detailed command execution can be seen with the VERBOSE=yes option. And in SCons, the counterpart for make clean is scons --clean or scons -c. Figures 8-13 and 8-14 are the screen captures for them. (BTW, you can always use scons -Q to remove the messages generated by SCons itself.)

Figure 8-13. SCons Verbose Build

Figure 8-14. *SCons clean*

To use the cross-compilers, some of the environment variables, such as CC and CXX, have to be replaced, as shown in the last section in Listing 8-15. Another useful option I can think of for SCons is --debug=explain, which will go through every step with detailed reasoning. It comes in very handy when you have doubts about the dependency decisions that SCons makes for you.

Ant and Maven

For the sake of completeness, I also want to briefly mention some XML-based build tools here. As you can tell from previous sections, makefile is a good match for building in the C/C++ language, and it is heavily dependent on shell scripts. However, when Java was brought to the fore, Java programmers soon found that the combination of a makefile and shell script was not good enough to achieve platform independence. So they came up with new build tools, such as Ant and Maven. My personal interaction with Maven was through Eclipse CDT, where I tried to make a new plugin for it at one point. What makes Maven different from others is that Maven will pull a lot of plugins all over the Internet and use them to build up the whole thing. In my eyes, such an approach works well for cross-platform software like Java, but it may not be a good fit for low-level software (firmware) that are used in the embedded systems.

Building Java programs is beyond the scope of this book. Inquisitive readers can find more detail of Ant and Maven in Ref [17] and Ref [18], respectively.

Embedded OS

The term "embedded system" runs a wide gamut. At low end, where the CPU has less horsepower and the storage space is compact, firmware runs on bare bones hardware. Hence the low-end systems can make do without any embedded OS. (One with an 8051 processor usually fits this profile.) For those systems, a super endless loop will suffice. Operations are carried out in serial fashion, plus some ISRs (Interrupt Service Routine) to handle asynchronous events.

However, as the system grows, so does the call for multi-threading and better support beyond a simple loop or HAL (hardware abstraction layer). This naturally brings embedded OS to the fore, as warranted by a more powerful CPU and larger memory.

Among all the embedded OS commercially available, embedded Linux is leading the pack[5]. Over the past decade, embedded Linux has been winning market share at the expense of its rivals. And it enjoys a huge developer community that actively refines it on daily basis. Consequently, it was chosen by this book as the epitome of what a typical embedded OS ought to be. As you can imagine, embedded OS is a pandemic subject. This section can only serve as a starting point on this subject.

Choose an Embedded OS

The choice of embedded OS comes from many dimensions. The following factors should play into your decision making.

Royalty and License Costs

Your profit margin will be eroded if the OS vendor charges a royalty. Although many vendors dropped royalty terms after the rise of the open source movement, it is still the first thing to look out for, as it will directly affect your bottom line.

The development costs have to be taken into account as well. Many vendors charge licensing fees for their development tools based on the number of seats. In addition, tech support costs should also be nailed down beforehand.

Open Source and GPL

Embedded OS with open source code has been gaining ground in the past decade. Open source brings more flexibility and less cost, along with a large developer community. Among them, Linux stands out as the bellwether, partly due to its popularity on the desktop PC.

However, adopting an open source kernel may also impose some unappealing duty on your code as well. Take the Linux kernel for example—it's licensed under the terms of the GPL, which stipulates that any derivative work is obligated to disclose its source code as well. So if you write a Linux device driver for your hardware, and you choose to link your driver into the kernel as a whole, that driver code is supposed to follow the open source terms. In practice though, many commercial products choose to implement their device drivers as a LKM (loadable kernel module) and ship them in binary form. The argument is that LKM is an attachment rather than a derivative work, and is thus not subject to GPL terms. Many companies, big and small, have been hiding their proprietary code under the LKM umbrella for years without any legal hiccups. Nevertheless, since it rides in a gray area, you might need to consult your legal department when it comes to the decision making.

---

[5]The renowned Android OS is also based on the Linux kernel.

Real Time, Interrupt Latency, and Throughput

Depending on the application scenario, your system may impose real-time requirements on the selection of OS. And there are two kinds of RTOSs (real-time OS) in the world: *hard real-time* and *soft real-time*.

An RTOS supporting hard real-time should have a predictable scheduling mechanism to guarantee that each task can be completed within a predetermined interval. No exception is taken. TDMA communication and pacemaker are two typical cases with hard real-time requirement. Besides the RTOS support, hard real-time should be considered at the system level. Each task should be carefully profiled. The CPU speed (normally measured in terms of MIPS) should be evaluated with meticulous number crunching to make sure there is enough computation power and a reasonable design margin.

On the other hand, under a soft RTOS, the majority of the tasks can be finished within the required interval, but a small amount of exceptions are also allowed in the interest of overall performance. A disk controller usually fits this profile, with most IOs being finished quickly, and extra latency on a small number of IOs poses no harm.

To achieve the optimal performance, designers normally have to strike a balance between interrupt latency and throughput. Those two benchmark indexes are closely tied to the RTOS beneath. Thus the maximum interrupt latency should be measured (or checked out through a datasheet) when evaluating the RTOS.

Preemptive and Non-Preemptive Kernels

OS kernels can also be categorized as preemptive and non-preemptive. Under preemptive kernels, a kernel task can be scheduled off CPU when a new one with higher priority becomes ready. However, for a non-preemptive kernel, the active task has to give up the CPU voluntarily before a new one can be scheduled, even when the new one might have higher priority. Tasks under non-preemptive kernels have to work in a very cooperative manner to achieve optimal performance.

Compared to preemptive kernel, non-preemptive ones may enjoy less overhead on rescheduling, but they also require better planning and collaboration among all tasks. Thus there are no carved-on-granite conclusions when choosing between a preemptive and non-preemptive kernel. The evaluation has to be conducted on a case-by-case basis.

Footprint Size

Most embedded OSs will be stored as a compressed image in Flash and expanded into RAM during bootstrapping. A small footprint will be a boon for cost-sensitive products. Thus the real estate expense, both in compressed and uncompressed form, should be taken into account during evaluation.

Hardware Requirements

Most embedded OSs will have a minimum hardware requirement in order in order to function properly. Your available choices are limited by such hard requirements. And those requirements are normally manifested as follows:

- Specific CPU architecture, such as ARM, MIPS, x86, etc.

- The existence of MMU.

- Minimum RAM size requirement.

- The existence of a particular peripherals, such as a graphic display device with certain dimensions and resolution, or a USB connection. As suggested in Ref [19], Android is a little picky in this regard.

Embedded Linux Development Flow

Despite the plethora of embedded OSs, the development flows of these systems are similar. This book will choose embedded Linux to demonstrate what a typical development flow ought to be.

Choose a Vendor

First off, you have to decide where to get your OS kernel. As for Linux, there are many ways to get an embedded Linux kernel for your target system:

- One way is to download the vanilla kernel from the Linux Kernel Archives (http://www.kernel.org). However, porting kernels to a customized system is not a small undertaking. You run the risk of hitting road blocks that are unforeseen at the beginning.

- In particular, the vanilla kernel may not fully support what you have in mind. For example, you might want to patch the kernel to reduce the interrupt latency for IO-intensive application scenarios. And there are vendors who provide their own flavors of kernel for various market segments. The good part of *not* doing it on your own is that you can focus your engineering resources on more important jobs and leave the kernel porting and patching to the hands of experts.

Development Machine Setup

In order to proceed efficiently, you need well-oiled development machines. By machine, what I really mean are servers, workstations, and storage boxes, as illustrated in Figure 8-15:

- *A storage server where your Unix home directory resides:* In a typical corporate IT environment, there are multiple Unix/Linux servers. You can have a SSO (Single Sign On) among those servers through services like NIS or LDAP. No matter what server you log into, your home directory looks exactly the same, because every server mounts its home directory to a centralized storage box. The storage box is normally a NAS (Network Attached Storage), with RAID (Redundant Array of Independent Disks) protection to survive single/multiple disk failure. Conceptually, you can view the storage box as a Linux server that exports NFS or Samba share. In other words, you should be able to map your Unix/Linux home directory to your Windows PC as a network drive.

- *Server for version control:* Your source code, along with all the schematic drawing and documents, is the most valuable asset of your company. You are supposed to treat them seriously by setting up a version control server (e.g., a subversion server) and protecting the data with RAID storage, as shown in Figure 8-15. Strong security policy should also be enforced when accessing this server.

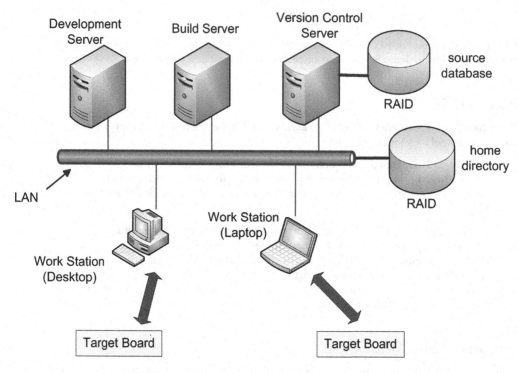

Figure 8-15. *Servers, workstations, and storage boxes*

- *Build server:* Obviously you have to turn your source code into binary images at some point. Although you can check out source code to your local drive and make intermittent builds on your workstation, it is hard to create a standard process from such a practice. Instead, an independent build server is recommended for the following reasons:

  1. It will be very hard to maintain exact the same toolchain across all the workstations. As time goes by, some team members may choose to upgrade the toolchain on their PCs[6] while others lag behind. A centralized build server can avoid such inconsistencies.

  2. There is no doubt that build servers can take the heavy load off workstations. On top of that, a build server is a first step toward CI (Continuous Integration, which will be discussed in later chapters). Basically, you can check out code from version control automatically and have nightly builds (or build on every commit). The build process can be streamlined to catch errors as early as possible.

---

[6]Here I take the liberty of assuming that PC is the standard service rifle for embedded developers. Apple fans, feel free to replace it with iMAC. :-)

- *Development server:* To develop with your embedded OS, embedded Linux in particular, you also need a dedicated server. The development server is responsible for the following jobs:

    1. *Provide a BOOTP/DHCP service to your development board*

        Most likely, your target system (or at least the development board) will have an Ethernet connection if you are working with embedded Linux. Although you can manually set up IP addresses for each individual board, it will be more convenient to let your development server do this for you automatically. However, your corporate IT network may also have a DHCP server in place. I suggest you coordinate with your IT department before you enable the BOOTP/DHCP service on your development server. To be on the safe side, it is better to isolate your lab network with your corporate IT network.

    2. *Provide a TFTP or FTP service to your development board*

        After your bootloader is successfully executed, it needs to locate your OS kernel image for second stage boot-up. During development, your kernel image can be loaded through RS-232, USB, or Ethernet[7], assuming burning Flash frequently is not a viable option. Although RS-232 is handy, its low throughput does not work well with large images. USB is a good alternative, but not every microprocessor has built-in support for booting from USB. In that sense, Ethernet is a more generic solution. After your board acquires an IP address, it can grab images (which might include kernel, driver, and apps altogether) through the TFTP protocol. (In this regard, some use FTP instead of TFTP.) Every time you have a new image, you can put it on your development server to be deployed to your target boards.

    3. *Provide an NFS service to your development board*

        The Linux kernel image is usually in a compressed form. After it is loaded into RAM, it will extract itself and start initialization. At some point, it will try to mount the root file system. During development, the root file system can be a NFS share exported by your development server, and accordingly your kernel image should be configured to mount the root file system through NFS. On the other hand, the root file system should reside on a Flash for your final product, and accordingly the production kernel image should be configured to use a Flash-based root file system.

        Using NFS mount gives developers more latitude, and it is normally preferred when things are still in their early stages.

---

[7]Of course, you can always use JTAG ICE. But ICE is an expensive piece of equipment, and it might be overkill if all you need is to load something into memory and run it.

- *Workstations*: In this regard, I recommend using a Windows-based PC for your development, even if you are targeting embedded Linux. Here is why:

  1. Most likely, you will need software for ICE or FPGA programming as well during the course of your embedded development. This type of software is usually Windows PC-based.

  2. For embedded Linux, you will need a good terminal emulator as your debug console. My personal experience is that the Windows-based ones, such as Tera Term, are easier to use than their Linux counterparts.

  3. I like to map my home directory to my PC, edit the code with my favorite Windows-based editors[8], and build the code on the server (telnet/ssh). Of course, this is more of a personal preference. But as mentioned earlier, you are supposed to do most of your builds on the server; that is what build server is set up for.

Prepare the Toolchain

To make your code into binary, you need a cross-compiler. More precisely, you need a toolchain that includes a cross-compiler, linker, assembler, debugger, and other tools (such as objdump, objcopy, etc.). The toolchain is supposed to be installed on the build server and maintained by designated personnel.

There are three main ways to get the toolchain:

- If you get your bootloader, OS kernel, or development tool from a vendor, the vendor will normally provide you with a pre-built toolchain. Having the blessing from your vendor can help you avoid many hidden roadblocks. So use your vendor's toolchain if you can get ahold of one.

- You can also download some pre-built toolchain for free from the Internet, such as the ELDK (Embedded Linux Development Kit, Ref [20]) from DENX Software Engineering, the pre-built GNU cross toolchain (Ref [21]) contributed by eCosCentric Limited, or the one from Linaro (Ref [29]).

- If you are using GNU toolchain, you can build the toolchain by yourself. But I suggest you do it only when you don't have any vendor supplied toolchain, or can't you find a free toolchain to match your host/target processor architecture.

Building a toolchain can be tedious, and it might also involve some special patches for your particular host/target processor architecture. If you choose uClibc as your C library, you can build your toolchain with BuildRoot (Ref [22]). Otherwise if you prefer glibc, you can get crosstool (Ref [23]) to help you build the toolchain.

Prepare the Bootloader

As mentioned in previous chapters, bootloader is the first program to be executed after the system powers on, and it is responsible for system initialization and for loading the OS kernel image. Because of this, embedded OS normally has its favored bootloaders. As for embedded Linux, bootloaders like RedBoot (Ref [24]) and U-Boot (Ref [25]) are popular among developers. After you get your toolchain ready, the next step is to port a bootloader to your target system.

---

[8]As you can imagine, I am no expert on emacs.

Although it is always possible to port these bootloaders on your own, a more practical approach is to start with a BSP (Board Support Package). When CPU vendors roll out their latest models, they, or some third-party companies, normally provide a demo board or development kit that has the CPU plus some peripherals and connectors. Along with the development kit, there will be a BSP that has the bootloader readily available. (Sometimes it also includes an OS kernel and device drivers.) Usually you can get the bootloader source from the board vendor and use that as your baseline. As far as bootloader is concerned, the delta between your target board and the development kit is most likely in memory address map and the peripherals.

Prepare the OS Kernel Image

Just like the bootloader, you can port the Linux kernel based on an existing BSP. (Assuming you can find one that meets your requirements.) Or you can get an OS vendor to do the porting for you. The nitty-gritty of Linux kernel is beyond the scope of this book. Refer to Ref [26] for more information.

As mentioned earlier, you can make two flavors of kernel images. One is intended for NFS-mounting to your development server; the other is a standalone kernel that mounts the root file system onto Flash. The former is convenient for development, and the image can be grabbed by the bootloader through TFTP. The latter is mainly for the final product.

Prepare the Device Driver

Demo board and development kit are good references for hardware design. Chances are that your target system may be a close cousin of those demo boards, which uses the same microprocessor. And the main delta in hardware design will most likely be in memory mapping and new peripherals.

For those new peripherals, you will have to write device drivers for them. Writing device driver requires deep knowledge of Linux kernel, and is thus beyond the scope of this book. Refer to Ref [26][27][28] for more information.

Your device driver can be linked to the kernel source, and thus become part of the kernel image. Or it can be a LKM (Loadable Kernel Module) that is loaded by shell script. As mentioned earlier, some companies use LKM to circumvent the GPL, and thus avoid disclosing their proprietary source code. However, since this is a legal gray area, consult your corporate lawyers for advice.

Prepare the Applications

As shown in previous chapters, anything above the OS and device driver can be called an application. And for applications, there are oodles of choices when it comes to their implementation. You can take the traditional approach by coding them in C/C++. Or you can use Java to create apps based on the virtual machine. You can even include a full-blown web server in your system if you have enough hardware resources. Due to the multitude of choices, the final decision has to be made on a case-by-case basis.

Prepare the Release Image

After you have all the previous tasks completed, you can start preparing the Flash image for your final product, which includes:

1. Configuring the kernel image to mount the root file system on Flash partitions.

2. Preparing the root file system. Instead of including the standard Unix/Linux command executables, you might consider using packages such as Busybox to significantly reduce the footprint size.

3. Packaging your device drivers and applications into Flash partitions. The partition can use read-only Flash file systems, such as CramFS.

4. Preparing the shell script to mount partitions, load device drivers, and start applications.

5. If necessary, preparing other Flash partitions, such as a JFFS2 partition for read/write.

6. Configuring the bootloader to load OS kernel images from Flash.

Finally, save the whole Flash image for release!

Summary

As the scale of your system grows, you need a formal build process to deploy your firmware consistently and promptly. The Make utility is a handy tool for this job. In fact, with the help of KConfig, you can even create for your build process a GUI-based configuration.

In addition to the Make utility, other build tools were also discussed and compared in this chapter. Among them, SCons stands out as a good alternative to Make.

Adopting embedded OS will be inevitable when multitasking becomes crucial to do your job. The second part of this chapter discussed the criteria for RTOS selection, as well as the general flow for embedded Linux development.

References

1. *Learning Perl on Win32 Systems.* Randal L. Schwartz, Erik Olson, Tom Christiansen, O'Reilly Media, August, 1997

2. *Learning Python, 3rd Edition.* Mark Lutz, O'Reilly Media, 2008

3. *Managing Projects with GNU Make, Third Edition.* Robert Mecklenburg, O'Reilly Media, November, 2004

4. "Recursive Make Considered Harmful." Peter Miller, *AUUGN Journal of AUUG* Inc., 1997

5. *Cx51 Compiler User's Guide,* "Optimizing C Compiler and Library Reference for Classic and Extended 8051 Microcontrollers," Keil Software Inc., September, 2011

6. "The Kernel Configuration and Build Process." Greg Kroah-Hartman, *Linux Journal,* Issue #109, May, 2003

7. kconfig-language.txt, (http://www.kernel.org/doc/Documentation/kbuild/kconfig-language.txt)

8. GNU Automake, version 1.11.1, David MacKenzie, Tom Tromey, Alexandre Duret-Lutz, Free Software Foundation, Inc., December 8, 2009

9. Autoconf, Creating Automatic Configuration Scripts, version 2.68, David MacKenzie, September, 2010

10. GNU M4, version 1.4.16, Rene Seindal, Francois Pinard, Gary V. Vaughan, and Eric Blake, Free Software Foundation, Inc., February, 28, 2011

11. CMake 2.8 Documentation (http://www.cmake.org), Kitware, Inc., Insight Software Consortium., 2009

12. JAM Product Information (http://www.perforce.com/documentation/jam)

13. Boost C++ Library (http://www.boost.org)

14. Boost.Build V2 User Manual, Vladimir Prus, 2009

15. SCons: A Software Construction Tool (http://www.scons.org)

16. *SCons User Guide 2.4.1,* Steven Knight and the SCons Development Team, 2015

17. *Ant: The Definitive Guide, 2nd Edition.* Steve Holzner, O'Reilly Media, April, 2005

18. Maven by Example, Sonatype Inc., 2011

19. *Embedded Android—Porting, Extending, and Customizing.* Karim Yaghmour, O'Reilly Media, October, 2011

20. ELDK 5.1 Documentation (`http://www.denx.de/wiki/ELDK-5/WebHome`)
21. Pre-built GNU cross toolchain (`ftp://ecos.sourceware.org/pub/ecos/gnutools/`), contributed by eCosCentric Limited
22. Buildroot: Making Embedded Linux easy (`http://buildroot.uclibc.org`)
23. Building and Testing gcc/glibc cross toolchains (`http://kegel.com/crosstool`)
24. RedBoot (`http://ecos.sourceware.org/redboot`)
25. Das U-Boot: The Universal Boot Loader (`http://www.denx.de/wiki/U-Boot/WebHome`)
26. *Linux Kernel Development, 3rd Edition.* Pearson Education, Inc., 2010
27. *Linux Device Drivers, 3rd Edition.* Jonathan Corbet, Alessandro Rubini, Greg Kroah-Hartman, O'Reilly Media, February, 2005
28. *Embedded Linux, Hardware, Software, and Interfacing.* Craig Hollabaugh, Ph.D., Pearson Education, 2002
29. Linaro (`https://www.linaro.org/`)

CHAPTER 9

Field-Programmable Gate Arrays

It was extremely important for us to always be at the leading edge and we felt that there's no way that we could build a fab.

—Bernard Vonderschmitt, Cofounder of Xilinx

Now comes the "hard" part. The next three chapters are all about hardware, which covers a wide gamut of topics like FPGA, SOPC, LCD, etc. This chapter begins with an overview of the embedded hardware in general, followed by detailed discussions about FPGA and IP protection.

Hardware Overview

The scope of hardware can be very broad, as broad as anything that has mass (or weight). In that sense, things like mechanical enclosures can be called hardware. But for this book, the scope of hardware covers only electronic components. As shown in Figure 9-1, the electronic components in most embedded systems can be divided into the following categories:

- Controller (a CPU, an FPGA, or a hybrid)

- Memory (including volatile and non-volatile)

- Peripherals

- Clock/reset and power management unit

Thus for all practical purposes, the focus of the next three chapters is on the following topics:

- FPGA and inter-FPGA connection

- IP protection

- SOPC (System on Programmable Chip)

- PMIC (Power Management IC)

- Clock and reset

- Switch (button) de-bounce

- Peripherals

© Changyi Gu 2016
C. Gu, *Building Embedded Systems*, DOI 10.1007/978-1-4842-1919-5_9

Peripherals can have a variety of forms. For display, it could be as crude as an RS-232 terminal, or as fancy as a colorful LCD display. For wireless communication, the RF component can be a single chip transceiver, or a combination of off-the-shelf components (mixer, synthesizer, etc.). Some of the peripherals shown in Figure 9-1 have already been discussed in earlier chapters. And LCD/touchscreen will be examined after we finish the topic of SOPC. (BTW, the list of peripherals in Figure 9-1 is by no means to be complete, as the whole component industry keeps charging forward relentlessly. But as for the control interface, most of them adopt the bus standards discussed earlier. So this book chooses to forgo them for the sake of brevity.)

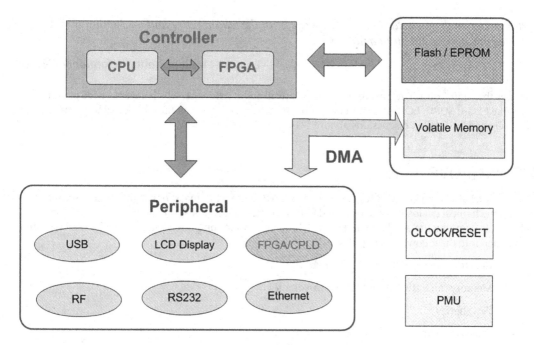

Figure 9-1. *Hardware overview*

FPGA

As its density keeps going up while the price is winding down, FPGA begins to gain a foothold in embedded systems. In fact, today's high device density has made it possible for FPGAs to have a soft core processor inside to completely replace an onboard CPU. This section will get started on FPGA. The SOPC (System on Programmable Chip) design approach will be heavily discussed in the next chapter.

CPU vs. FPGA

As demonstrated in Figure 9-1, all embedded systems have at least one controller as their brain. There are two ways to make this brain: software approach (by CPU) or hardware approach (Digital Logic through FPGA/ASIC). The software approach has been discussed in detail in previous chapters. For most single-core

CPUs, instructions are executed in serial fashion. (VLIW might be an exception.) To achieve multitasking, tasks are carried out concurrently. As shown in Figure 9-2, at any single moment, only one task is active. The multitasking is fulfilled by context switching among those tasks, which means there is no true parallelism for this software approach.

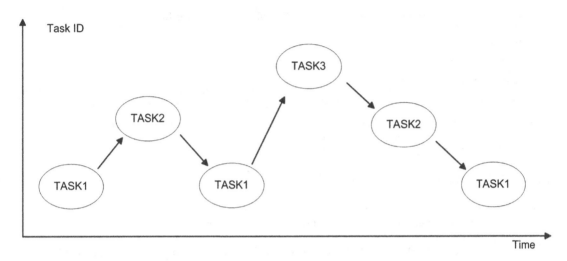

Figure 9-2. *Multitasking, concurrent execution*

In software, complicated control flow can be achieved through programming in a high-level language, which is a boon for design and maintenance. However, the software approach also has its share of shortcomings:

- Due to the absence of true parallelism, multitasking has to be accomplished by context switching, and such switching is done asynchronously. From the design standpoint, asynchronism always poses a challenge. The biggest headache for any asynchronous design is the lack of good methodology to verify the design[1]. I guess that's part of the reason why software is often plagued with bugs.

- Another challenge for software approach is to attain real time. For a single-core CPU, the main avenue for more MIPS is to increase its clock frequency. However, due to the constraint of semiconductor process, clock frequency cannot be increased indefinitely. In addition, higher frequency will lead to more power consumption, which may not be realistic for battery-powered devices.

As an alternative approach, the controller can also be implemented by digital logic, which can be physically realized by FPGA or ASIC. Compared to the software approach, digital logic has its merits:

- True parallelism can be implemented effortlessly through digital logic. Its clock frequency can be much lower than its CPU counterpart, thanks to its parallel architecture.

- Digital logic is primarily done by synchronous design. The simulation and verification for synchronous design is much more mature and reliable than its asynchronous counterpart. Thus real time can be attained with less effort in digital logic.

---

[1]So far there are still no systematic approaches to reliably verify or prove the correctness of asynchronous design.

As summarized in Table 9-1, each approach has its pros and cons. The rest of this chapter focuses on the FPGA solution.

Table 9-1. *Software (CPU) vs. Digital Logic (FPGA/ASIC)*

	Software (CPU)	Digital Logic (FPGA/ASIC)
Clock frequency	High	Low
Parallel architecture	No	Yes
Power consumption	High	Low
Achieve real time	Relatively hard, less reliable	Relatively easy, more reliable
Design and maintain	High-level programming language (C, C++, etc.)	Hardware description language (Verilog/VHDL, etc.)

Verilog vs. VHDL

Digital logics are primarily designed by an HDL (Hardware Description Language). As with the case of software programming languages, there exists more than one HDL. Fortunately, only two of them are chiefly used in the industry: Verilog and VHDL.

The choice between Verilog and VHDL may trigger another holy war, which I have no intention in getting involved in. However, since I had the chance to work with both languages, I would like to offer my personal opinions from a designer's perspective.

Under most circumstances, the decision to choose between Verilog and VHDL had already been made a long time ago when your design team was initially formed. Since then libraries and code bases were created with that particular language. And new designs will always gravitate toward the same language for the sake of code reuse. Historically, VHDL is widely used by a great number of defense contractors and European companies, while Verilog is popular among commercial design houses in North America.

However, if you were able to make the call, I would recommend Verilog instead of VHDL for the following reasons:

1. VHDL is a strong typed language while Verilog is not. To use programming languages as an analogy, VHDL is to Verilog what Pascal is to C. However, since VHDL and Verilog are both intended to describe hardware, every value eventually boils down to ones and zeros. The strong type offers little help besides making things awkward.

2. The evolution of VHDL is lagging behind that of Verilog. Many useful features of Verilog 2001 can only find their counterparts in VHDL 2008. To name a few, here are two conspicuous examples:

    •   For verification, Verilog testbench supports hierarchical names to tap into the signals buried deep down beneath. And such a feature only gets picked up as late as VHDL 2008 through external names.

    •   For the design of combinational logic, Verilog 2001 supports wildcards in the sensitivity list. A similar feature is only offered by VHDL in its 2008 standard, as demonstrated in Listing 9-1. But even so, it took many EDA vendors a long while to fully support this feature in their simulation tools[2].

---

[2]For example, `process(all)` is not supported by Modelsim until version 10.0.

Listing 9-1. Wildcards in Sensitivity List

Verilog 2001

```
always (*) begin

    a <= b & c;

end
```

VHDL 2008

```
process(all)
begin
  a <= b and c;
end process;
```

3. Initially both VHDL and Verilog are weak on verification, so they have to rely on other languages, such as Vera or e for more advanced verification. But the introduction of SystemVerilog has greatly changed the landscape. As a superset of Verilog, System Verilog is rich in features to enhance verification. In this regard, VHDL has no counter-offers. In addition, SystemVerilog has also borrowed pages from VHDL's playbooks to make it handy and attractive to designers. Thus for newly started designs, I strongly recommend SystemVerilog over pure Verilog, let alone VHDL.

4. For gate-level simulation, Verilog holds a performance advantage over VHDL, because it has built-in primitives to describe gate-level components. VHDL has to rely on VITAL library in this regard. In fact, most third-party simulation libraries are in Verilog. So if your design is in VHDL, you have to acquire a mixed language license in addition to your existing VHDL license when you simulate a design that contains a third-party library. As you know, EDA licenses are extremely pricy.

The rest of the book presents designs in Verilog and SystemVerilog.

Controller and Datapath

The nonstop refinement of semiconductor process has made it possible for FPGAs to have great density. Today, complicated designs with millions of gates can be hold in a single FPGA chip. And large-scale design calls for a systematic design approach, which usually means divide-and-conquer.

As suggested in Ref [4], most digital designs, no matter how complicated they are, are made of two major components: the *controller* and *datapath*. Seen from the top level, the controller is the brain of the design, which usually has a FSM (Finite State Machine) that sends out all the commands (control signals). On the other hand, the datapath is a medley of various function units. Each function unit will take order from the controller and carry out certain operations, such as counter, ALU, shift registers, etc. For large design, the function unit could have its own controller and datapath. As the design scale goes up, more design layers can be in place, and the second level function unit can also form its own controller and datapath, as illustrated in Figure 9-3. Such a divide-and-conquer strategy can be executed iteratively to the last design node, where the function unit is manageable without the presence of FSM.

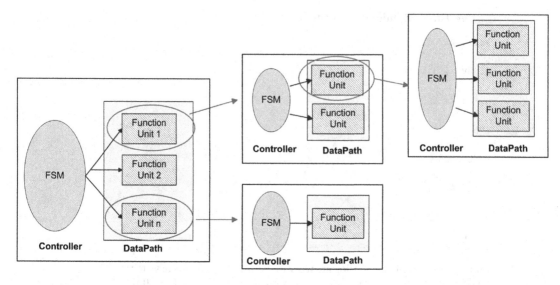

Figure 9-3. *Divide and conquer*

Implement FSM in FPGA

Since the intelligence of digital design is mostly reflected on the construction of FSM, it is imperative that FSM gets implemented efficiently in the hardware. Conceptually, a FSM can be represented by either a state diagram or a state transition table, as exemplified in previous chapters. Its software implementation has been examined earlier as well. However, from the standpoint of digital design, a FSM is made of two parts: the registers to keep the current state (state register) and the combinational logic that determines the state transition and output.

As illustrated in Figure 9-4, the state register is supposed to get its next state from a big trunk of combinational logic. And here comes an interesting question: If you have an FPGA design with n states in the state diagram, how many flip-flops would you have for the state register? The conditioned response seems to be "ceiling($\log_2 n$)" since it reduces the number of flip-flops to the minimum. However, reducing the number of flip-flops may not be optimal for FPGA implementation, because the traditional FPGA architecture is often made of LUTs (lookup tables) and flip-flops. Reducing the number of flip-flops usually means bigger trunk of combinational logic, since more effort has to be spent on state decoding. In turn, the resulting larger combinational logic has to be partitioned and spread out among multiple logic blocks inside FPGA. However, the biggest problem with this approach is that more routing resources are needed to interconnect those logic blocks. The routing resource is the most precious resource for FPGA. Heavy consumption of routing resources will lead to larger area and longer delay. On the other hand, taking up some extra flip-flops may not be such a big deal for FPGA, since flip-flops are abundant in each logic block and interconnection among them can be handled chiefly at local scale.

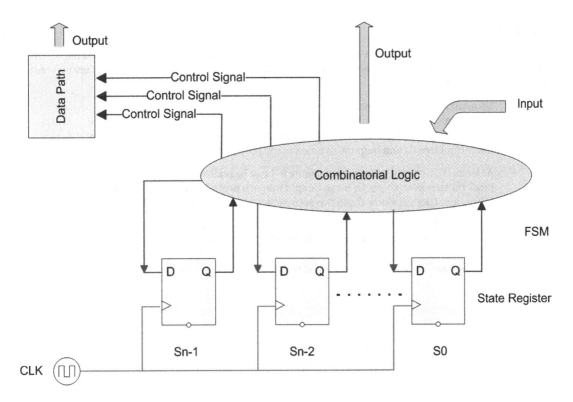

Figure 9-4. *Implementing FSM in FPGA*

Thus, FSMs are usually implemented in FPGA with one hot encoding instead of binary encoding, which means for a FSM of n states, n flip-flops are used for state register. At any given moment, one and only one of the n flip-flops can be set to high (Ref [6][8]). The one-hot encoding can be written directly into your HDL code (as demonstrated in later sections), or it can be set by using proper synthesis options.

Another note worth mentioning for Figure 9-4 is the registering of control signals and outputs. As you may know, combinational logic is the enemy of timing. If you have trouble closing timing, you might have to register your control signals or outputs, or even put deeper pipelines for the combinational logics in Figure 9-4. (Some HDL coding style will naturally do this for you, although it can cause extra cycles of delay.) In addition, if your output is heading for another clock domain, it is absolutely necessary to register your output, as suggested in later sections.

HDL Template for FSM

There are many ways to write Verilog/VHDL code for FSM. Sometimes, it is just a matter of personal preference. Examples with various implementation styles are abundant in Ref [4][5][7][9]. And my two cents on this subject are as follows:

- No matter what implementation style you choose, the RTL net list after synthesis might bear striking similarities. And most likely, the resulting net list will be a close cousin to what's in Figure 9-4. (Some implementation styles may carry registers on the control signals and outputs, as noted in the previous section.)

- As far as HDL coding is concerned, Figure 9-4 can be realized with one procedural block that has everything. Or it can be split into two procedural blocks. For the sake of argument, let's see how those two approaches fair on a simplified FSM.

  Figure 9-5 shows a Moore FSM with only two states: IDLE and RUN. The switch between these two states is triggered by the ctl_start and ctl_stop inputs, and the output active will be set to 1 if the current state is RUN. The data path that accompanies the FSM is just a counter, which counts only under the RUN state.

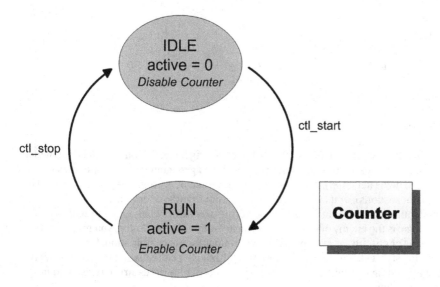

Figure 9-5. *Two-state FSM*

Given the state diagram in Figure 9-5, the corresponding single-procedural-block realization is demonstrated in Listing 9-2[3]. A close look at Listing 9-2 reveals the following deficiencies:

a. The output `active` is not totally in sync with the state change. There is a latency of one clock cycle from the state change to the output change, as shown in Figure 9-6. Such latency is introduced because output is dependent on the state change, while they both work under sequential logic.

b. If such latency bothers you, you can put the output generation into a separate `always` statement of combinational logic, as illustrated in Listing 9-3. However, the second `always` statement also needs to enumerate through all the possible states. If the state diagram is a big one with lots of states, such an approach can be cumbersome.

c. The data path in Listing 9-2 is directly coupled with the FSM states. That means, there are no explicit control signals generated from the FSM to the data path. Any redesign or improvement of the state diagram will directly impact the data path as well. Using explicit control signals can alleviate such problems since conceptually the data path is relatively independent of the FSM. However, if everything resides in only one `always` statement for FSM, all control signals will experience the same one-clock-cycle latency as the outputs.

To handle these dilemmas, you can adopt a new approach with two procedural blocks. One block is in sequential logic while the other is in combinational logic. However, different from Listings 9-2 and 9-3, the sequential logic here does not deal with the actual states. Instead, it only updates the state register with new states generated from the combinational procedural block. And the new state, along with all controls signals, is solely determined by the combinational block, as illustrated in Listing 9-4. (Listing 9-4 is based on the template in Listing 9-5.)

The good thing about Listings 9-4 and 9-5 is that there is only one place to enumerate all the possible states, and control signals are explicit. In addition, the latency you saw in Listing 9-2 no longer exists (Figure 9-7). But if I have to say something negative about this approach, it is that everything is handled in a big trunk of combinational logic. You might have to add extra registers on control signals or implement deeper pipelines for better timing closure. But you only do this on an as-needed basis.

Listing 9-2. FSM with One Procedural Block

Verilog 2001

```
module one_always_FSM(
//========== INPUT ==========

    input wire clk,
    input wire reset_n,
    input wire ctl_start,
    input wire ctl_stop,
```

---

[3]For the sake of one-hot encoding, a reversed case statement is adopted where 1'b1 is put in the case expression. A reverse case statement is good for synthesis because only the active FF is compared instead of the whole vector (Ref [10]).

```
//========== OUTPUT ==========
    output reg [7:0] counter_out,
    output reg active
//========== IN/OUT ==========

);

    localparam STATE_IDLE = 0, STATE_RUN = 1;

    // state register
    reg [1:0] state;

    // FSM
    always @ (posedge clk or negedge reset_n) begin
        if (!reset_n) begin

            state <= (1 << STATE_IDLE);
            active <= 1'b0;

        end else begin

            case (1'b1)
                state[STATE_IDLE]: begin
                    active <= 1'b0;

                    if (ctl_start) begin
                        state <= (1 << STATE_RUN);
                    end else begin
                        state <= (1 << STATE_IDLE);
                    end
                end

                state[STATE_RUN]: begin

                    active <= 1'b1;

                    if (ctl_stop) begin
                        state <= (1 << STATE_IDLE);
                    end else begin
                        state <= (1 << STATE_RUN);
                    end
                end

                default: begin
                    state <= (1 << STATE_IDLE);
                    active <= 1'b0;
                end

            endcase
        end
    end // end of always
```

```
    // Data Path
    always @(posedge clk or negedge reset_n) begin
        if (!reset_n) begin
            counter_out <= 0;
        end else begin
            if (state[STATE_RUN]) begin
                counter_out <= counter_out + 1;
            end
        end
    end // end of always

endmodule // one_always_FSM
```

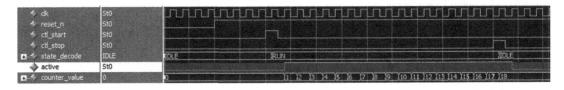

Figure 9-6. *Timing diagram of Listing 9-2*

Listing 9-3. Combinational Logic to Remove Latency

Verilog 2001

```
    always @(*) begin
        case (1'b1)
            state[STATE_IDLE]: begin
                active = 1'b0;
            end

            state[STATE_RUN]: begin
                active = 1'b1;
            end

            default: begin
                active = 1'b0;
            end
        endcase
    end // end of always
```

Listing 9-4. FSM with Two Procedural Blocks

Verilog 2001

```
module two_always_FSM(
//========== INPUT ==========

    input wire clk,
    input wire reset_n,
```

```verilog
    input wire ctl_start,
    input wire ctl_stop,
//========== OUTPUT ==========
    output reg [7:0] counter_out,
    output reg active
//========== IN/OUT ==========

);

    localparam STATE_IDLE = 0, STATE_RUN = 1;

    // state register
    reg [1:0] current_state, next_state;
    reg ctl_enable_counter;

    // FSM
    always @(posedge clk or negedge reset_n) begin
        if (!reset_n) begin
            current_state <= (1 << STATE_IDLE);
        end else begin
            current_state <= next_state;
        end
    end // end of always

    always @(*) begin
        next_state = 0;

        ctl_enable_counter = 0;
        active = 0;

        case (1'b1)
            current_state[STATE_IDLE]: begin
                active = 1'b0;

                if (ctl_start) begin
                    next_state[STATE_RUN] = 1'b1;
                end else begin
                    next_state[STATE_IDLE] = 1'b1;
                end
            end

            current_state[STATE_RUN]: begin
                active = 1'b1;
                ctl_enable_counter = 1'b1;

                if (ctl_stop) begin
                    next_state[STATE_IDLE] = 1'b1;
                end else begin
                    next_state[STATE_RUN] = 1'b1;
                end
            end
```

```
        default: begin
            next_state[STATE_IDLE] = 1'b1;
        end
    endcase

end // end of always

// Data Path
always @(posedge clk or negedge reset_n) begin
    if (!reset_n) begin
        counter_out <= 0;
    end else begin
        if (ctl_enable_counter) begin
            counter_out <= counter_out + 1;
        end
    end
end // end of always

endmodule // two_always_FSM
```

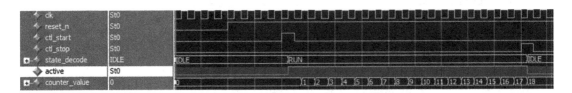

Figure 9-7. *Timing diagram of Listing 9-4*

- In the early days, the coding style in Listing 9-5 was not favored by many designers, due to the big sensitivity list required for combinational logic. At that time, HDLs were not smart enough to support wildcard[4] in sensitivity, and the sensitivity list had to be constructed manually, which was error-prone and tedious. But this is no longer the case. Consequently, I recommend the template in Listing 9-5 for your FSM construction on FPGA.

- With the acknowledgement of SystemVerilog, designers are gradually adopting it as the main design HDL. So you might consider using the template in Listing 9-6 for new designs. However, be advised that some features of SystemVerilog are not fully supported by certain synthesizers at this moment. (For example, enumerator methods are not supported by all synthesizers.)

---

[4]In Verilog 2001, wildcard is supported by using `always @(*)`. In VHDL 2008, it is supported by using `process(all)`.

Listing 9-5. Recommended FSM Template in Verilog 2001

Verilog 2001

```verilog
`default_nettype none

module MODULE_NAME (
//========== INPUT ==========

    input wire clk,
    input wire reset_n,
    input wire ...
    input wire ...
//========== OUTPUT ==========
    output ...
//========== IN/OUT ==========
    ...
);

    /* TO DO : define FSM states here */
    localparam STATE_INIT = 0, STATE_S1 = 1 ... STATE_Sn-1 = n;

    // current state and next state
    reg [n - 1:0] current_state, next_state;

    // State Register
    always @(posedge clk or negedge reset_n) begin
        if (!reset_n) begin
            current_state <= (1 << STATE_INIT);
        end else begin
            current_state <= next_state;
        end
    end // end of always

    // State Transition
    always @(*) begin
        next_state = 0;

    /* TO DO : all control signals and outputs default to zero */

        case (1'b1)
            current_state[STATE_INIT]: begin
              /* TO DO : fill in the code */
            end

            current_state[STATE_S1]: begin
              /* TO DO : fill in the code */
            end
```

```
        /* TO DO : construct the whole FSM here */

        default: begin
            next_state[STATE_INIT] = 1'b1;
        end
    endcase

end // end of always

/* TO DO : put data path here */

endmodule

`default_nettype wire
```

Listing 9-6. Recommended FSM Template in SystemVerilog

System Verilog

```
`default_nettype none

module MODULE_NAME (
//========== INPUT ==========

    input wire clk,
    input wire reset_n,
    input wire ...
    input wire ...
//========== OUTPUT ==========
    output ...
//========== IN/OUT ==========
    ...
);

    /* TO DO : enumerate FSM states here */
    enum {S_IDLE, S_RUN, S2, S3...} states;

    // total number of states
    localparam FSM_NUM_OF_STATES = states.num();

    // current state and next state
    var logic [FSM_NUM_OF_STATES - 1:0] current_state, next_state;

    // State Register
    always @(posedge clk or negedge reset_n) begin
        if (!reset_n) begin
            current_state <= (FSM_NUM_OF_STATES)'(1 << S_IDLE);
        end else begin
            current_state <= next_state;
        end
    end // end of always
```

```
    // State Transition
    always_comb begin

        next_state = 0;

     /* TO DO : all control signals and outputs default to zero */

        case (1'b1)
            current_state[S_IDLE]: begin
              /* TO DO : fill in the code */
            end

            current_state[S_RUN]: begin
              /* TO DO : fill in the code */
            end

            /* TO DO : construct the whole FSM here */

            default: begin
                next_state[S_IDLE] = 1'b1;
            end
        endcase

    end // end of always

    /* TO DO : put data path here */

endmodule

`default_nettype wire
```

Bi-Directional Bus

It might be necessary for you to support a SRAM-like interface (close to the one mentioned in previous chapters), for the purpose of driving a LCD module through parallel bus, or when interacting with microcontrollers. Consequently, you will find it essential to deal with bi-directional data bus. Implementing a bi-directional bus usually involves a tri-state buffer.

A typical bi-directional bus can be done in the way shown in Figure 9-8. In Figure 9-8, a SRAM-like interface is implemented by FPGA to talk to an external microcontroller. For the sake of simplicity, address bus and chip enable are not taken into consideration. The corresponding data bus can be written in Verilog as demonstrated in Listing 9-7. The tri-state buffer is controlled by the OE* pin. To avoid driving the data bus during the write cycle, it will output high-impedance when OE* is low. (The design demonstrated here assumes Intel 8080 bus type is used.)

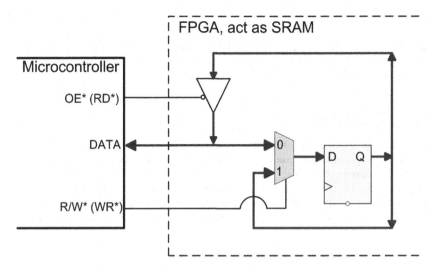

Figure 9-8. *FPGA acting as SRAM, with a bi-directional data bus*

Listing 9-7. Implement SRAM Bi-Directional Bus in Verilog

```verilog
module FPGA_SRAM (
//========== INPUT ==========

    input wire clk,
    input wire reset_n,
    input wire oe_n,
    input wire rw_n,
//========== OUTPUT ==========
    ...
//========== IN/OUT ==========
    inout wire [7 : 0] data
    ...
);

    reg [7 : 0]    sram_data;

    assign data = ? oe_n : sram_data : 8'bZ;

     ...

    // SRAM data
    always @(posedge clk or negedge reset_n) begin
        if (!reset_n) begin
            sram_data <= 0;
        end if (!rw_n) begin
            sram_data <= data;
    end // end of always
```

Clock

Just like human beings live due to a regular heartbeat, all digital circuits (including microprocessors) rely on clocks to proceed and synchronize. In EE's lingo, it means we need something that can oscillate stably at a certain frequency. At the circuit board level, it can be done in the following fashion.

Use External Crystal for Oscillation

Microcontrollers often use external crystals to make the oscillation. As shown in Figure 9-9, external to the microcontroller are one resistor, two capacitors, and one crystal with the designated frequency. The resistor provides bias needed for the internal inverter/amplifier. The two capacitors—Cx and Cy—and the crystal Y provide a 180-degree phase shift at the resonance frequency. And the internal inverter/amplifier gives another 180-degree shift, which makes total phase shift 360 degrees. Thanks to the amplifier, the total loop gain will be more than 1. Altogether, the loop gain and phase shift of 360 degrees form the basis for oscillation.

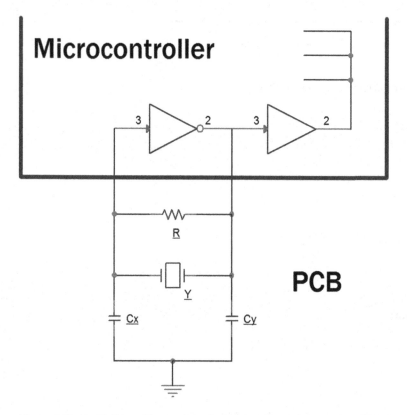

Figure 9-9. *Oscillation with an external crystal*

For all practical purposes, microcontroller's datasheet usually will provide the recommended value for Cx, Cy, R, and Y. And those components should stay as close to the microcontroller as possible to reduce the inductance introduced by PCB trace or component leads. Also note that it is never a good idea to take out Cx and Cy in favor of lower BOM cost, although engineers can sometimes observe successful oscillation without the presence of Cx and Cy. Most likely, it is the parasitic capacitance that acts as the role of Cx and Cy in that case. Relying on parasitic capacitance often leads to low yield during mass production.

The component cost for using external crystal is relatively low, and so is its accuracy. And it could also become both the source and the victim of EMI interference.

Canned Oscillator

To achieve better frequency stability and reduce interference, the whole circuit shown in Figure 9-9 can be sealed into one package. Oftentimes, the package is a metal case in DIP package[5]. And people call it canned oscillator. It comes with four pins: the power, the ground, the output, and a NC (no connection) pin for mechanical durability.

Using an oscillator can reduce the number of components and save PCB real estate. When used with FPGA, the output of oscillator should go directly to FPGA's dedicated clock input pin, followed by a PLL inside the FPGA to generate clocks at the desired frequency.

TCXO, OCXO, and VCXO

TCXO stands for Temperature Compensated Crystal Oscillator. OCXO stands for Oven Controlled Crystal Oscillator. VCXO stands for Voltage Controlled Crystal Oscillator.

A typical canned oscillator may achieve frequency stability as good as +/- 20ppm (parts per million). When more frequency stability is preferred over a wide range of temperatures, a temperature compensation circuit has to be designed into the oscillator. And this is where TCXO comes into the picture.

A good TCXO might give you a frequency stability of +/- 0.5 ppm over -20°C to +70°C. If more stringent accuracy is desired, an OCXO can be used instead. OCXO is often bigger than its TCXO counterparts, because the whole compensation circuit is placed inside a tightly sealed container (as the name "oven" suggests). In addition, a temperature sensor is also placed inside the oven to regulate the temperature to a much higher than normal range, like between +70°C and +80°C (Ref [13]), where the crystal's temperature coefficient is the smallest. With the constant temperature and small temperature coefficient, typical OCXOs can achieve frequency stability in the order of ppb (parts per billion) instead of ppm.

Notwithstanding manufacturers best efforts, no two oscillators are exactly the same due to aging, vibration, humidity, ambient temperature, etc. In telecom, both the transmitter and receiver can tune each other's local oscillator to a standard reference, such as GPS. In such a case, a VCXO (more often than not, it will be a TCVCXO) is used. As its name suggests, a VCXO's frequency can be tuned with an external voltage. Oftentimes, an external DAC controlled by FPGA or microcontroller is used to set the voltage for the VCXO.

PLL (Phase Lock Loop)

Inevitably, there will be phase noise associated with any oscillator in the physical world. And it will be reflected in the time domain as jitters. To reduce jitter/phase noise, a PLL is often used. PLL usually includes a loop filter and a VCXO to track the reference clock and reduce the jitter/phase noise. And it can also multiply the reference clock when a counter is used in its feedback path. For discrete PLLs, they can often be programmed through I2C or SPI bus. And these days, most FPGAs and microcontrollers will have PLLs integrated inside. The PLL can be used as a reliable clock source, and its "locked" indicator pin can also be used for asynchronous resets.

---

[5]Note that metal case/DIP is not the only package available. The oscillator could be in any package, such as a plastic case with a surface mount package.

RC Oscillator

On the other hand, a low-cost, low-accuracy alternative is to use pure electronic components for oscillation, as illustrated in Figure 9-10, where the two inverters will each provide 180-degree phase shift. Such kind of oscillator is often called RC oscillator and has a frequency stability of 1% or more. They are often integrated into many low-end FPGAs or microcontrollers to reduce BOM cost.

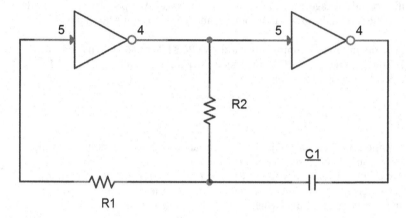

Figure 9-10. *RC oscillator*

Reset

It is often desired to be able to put the whole system into a known initial state. That is what reset is for. There are two kinds of reset in the digital design: synchronous reset and asynchronous reset. The synchronous reset has to rely on the clock being available to work while the asynchronous reset does not. They can usually be written as shown in Listing 9-8.

Listing 9-8. Synchronous Reset and Asynchronous Reset

Verilog 2001

```
    always @ (posedge clk or negedge reset_n) begin
        if (!reset_n) begin
            q <= 0;
        end else begin
            if (sync_reset) begin
                q <= 0;
            end else begin
                q <= d;
            end
        end
    end // end of always
```

System Verilog

```
always_ff @ (posedge clk, negedge reset_n) begin
    if (!reset_n) begin
        q <= 0;
    end else begin
        if (sync_reset) begin
            q <= 0;
        end else begin
            q <= d;
        end
    end
end // end of always
```

After synthesis, the synchronous reset will become nothing more than a mux control, as illustrated in Figure 9-11. The output of the mux goes directly into the input of the register. On the other hand, each register has a dedicated reset pin for asynchronous reset, and asynchronous reset can be done without the presence of a valid clock.

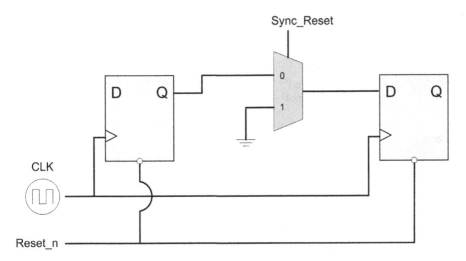

Figure 9-11. *Schematic for synchronous reset and asynchronous reset*

As far as FPGA design is concerned, the recommended coding style for choosing between synchronous reset and asynchronous reset is really vendor dependent. The drawback of a synchronous design is that it has to insert some combinational logic into the data path, which could affect timing negatively. And it only works when the clock is available.

If you decide to go with synchronous reset, and if the sync-reset signal has to drive a good number of flip-flops, my recommendation is to let the sync-reset go through a shift register (i.e., using flip-flops to delay it by a few clock cycles.) and use its delay version to drive the whole circuit. Meanwhile, turn on the physical synthesis to enable re-timing and register duplication. This will help to close the timing if a large fan-out is needed.

Nevertheless, things will get trickier when asynchronous reset is adopted. For FPGA design, the origin of the asynchronous reset could come external to the FPGA, such as from a pushbutton on the PCB. In that case, button de-bouncing needs to be implemented. Button de-bounce is discussed in later sections.

For the time being, let's assume button de-bounce has been taken care of. The asynchronous reset input usually has two places to go inside an FPGA:

- It can be used to reset the built-in PLL asynchronously.

- It can be used as the master reset to asynchronously reset all the relevant flip-flops inside FPGA.

For case 1, as illustrated in Figure 9-12, FPGA's built-in PLL will try to reduce the jitter on the external oscillator and provide a frequency multiplier at the same time. Its "locked" pin is the status output of the PLL, and it can be used as the master asynchronous reset for the rest of the logic circuit.

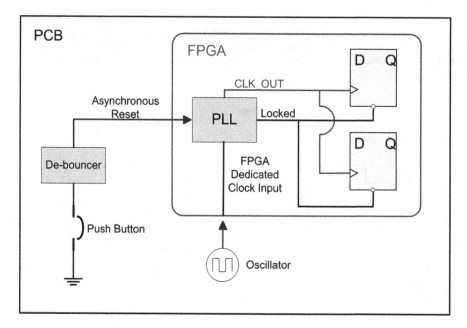

Figure 9-12. *Asynchronous reset using PLL locked pin*

On the other hand, for case 2 where no PLL is involved, a synchronizer is needed to make sure there is no meta-stability when the asynchronous reset is removed (Ref [14][15]). Without the synchronizer, the removal of the reset asynchronously could introduce two problems:

- The removal could happen very close to a clock rising edge, which will cause reset recovery time violation (Ref [15]). This in turn will introduce meta-stability.

- The delay from the reset original to each flip-flop varies. Due to the possible meta-stability mentioned in (1), flip-flops may not come out of the reset state on the same clock if reset is removed asynchronously.

Thus a synchronizer like the one shown in Figure 9-13 should be adopted to make the "removal of asynchronous reset" synchronous to the clock edge. Note that for the second flip-flop in Figure 9-13, both its input and its output are low when reset is removed. So there is no meta-stability introduced (Ref [15]).

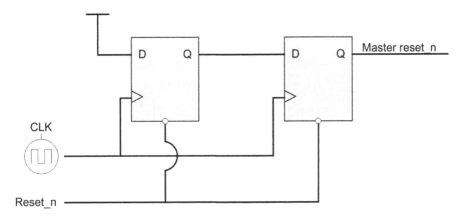

Figure 9-13. *Synchronizer for asynchronous reset*

CDC (Clock Domain Crossing)

As mentioned earlier, digital designers should practice synchronous design as much as possible. However, there could be multiple clock domains in reality, which make it necessary for signals to cross from one clock domain to another. CDC (Clock Domain Crossing) is asynchronous by its nature, and it is hard to catch asynchronous design flaws through simulation only. For reliable CDC design, designers could use the following guidelines.

Single Bit, Level Signal

The most basic case for CDC is single bit, level signal. For which, the common practice is to use two flip-flops in a row as a synchronizer (Ref [1]), as illustrated in Figure 9-14. The only caveat is that, before the signal leaves the first clock domain, it should go through a flip-flop (hash marked in Figure 9-14) to remove any possible glitches. For synchronous circuit, glitches are not a big concern. However, since these glitches are asynchronous to the second clock domain, they might be picked up by the two-FF synchronizer as a valid signal, which leads to unknown consequences. Thus it is vital not to keep combinational logic between the two clock domains.

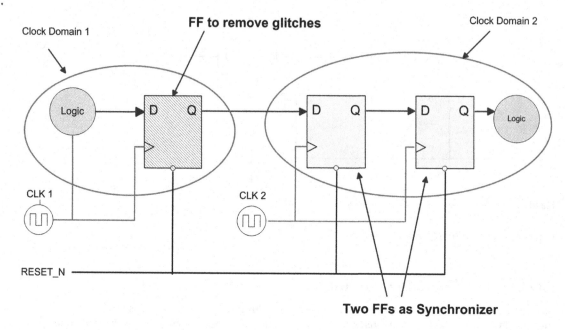

Figure 9-14. *Using two flip-flops as a synchronizer*

SIDE NOTES

My first job as a digital designer was to design a transmit chain in FPGA. One end of the transmit chain went to a WCDMA RF transceiver, which worked on a clock that was multiple of 3.84MHz. The other end of the transmit chain was connected to a microcontroller, which worked on its bus clock. The microcontroller turned on the transmitter by writing a control register that was in the bus clock domain. Those control signals had to cross into the RF clock domain to control the RF transceiver.

I screwed things up by accidently introducing an AND-gate on the path between the two red circles in Figure 9-14. The transmitter was working okay except it would go off for no reason every half an hour, which was followed by hair loss and insomnia for two consecutive days.

Single Bit, Pulse Signal

Sometimes instead of sending a level signal, you'll want to send a pulse across the clock domain. The pulse should be active for only one cycle in both clock domains. For this, a circuit called a *toggle synchronizer* is often used in practice (Ref [2]). As illustrated in Figure 9-15, the idea is to convert the pulse into a level signal (toggle), send it into the second clock domain through the same two-FF synchronizer in Figure 9-14, use edge detector to catch its rising edge or falling edge, and regenerate a single cycle pulse in the second clock domain.

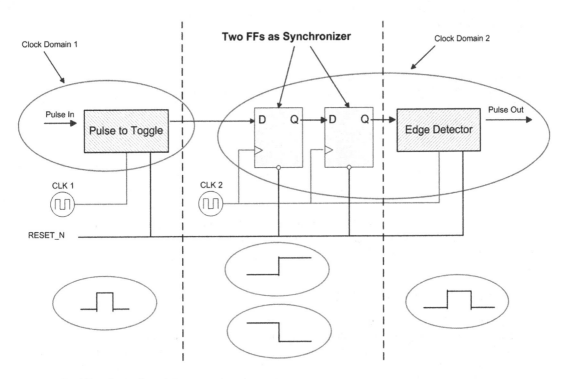

Figure 9-15. *Toggle synchronizer*

In Figure 9-15, the Pulse_to_Toggle converter can be implemented with one flip-flop toggling on the active polarity. Its Verilog/VHDL code can be found in Listing 9-9, and the code for the Edge Detector can be found in Listing 9-10.

Listing 9-9. Pulse to Toggle

Verilog 2001

```
parameter POLARITY = 1;

always @(posedge clk or negedge reset_n) begin
    if (!reset_n) begin
        toggle_out <= 0;
    end else begin
        if (data_in == POLARITY) begin
            toggle_out <= ~toggle_out;
        end
    end
end
```

Listing 9-10. Edge Detector

Verilog 2001

```verilog
always @(posedge clk or negedge reset_n) begin
    if (!reset_n) begin
            data_in_d1 <= 0;
    end else begin
            data_in_d1 <= data_in;
    end
end

assign pulse_out = data_in ^ data_in_d1;
```

Toggle Synchronizer is widely used in digital design to send data enabled strobes across the clock domain, or to implement handshaking protocol. Make sure there are enough gaps between every two consecutive pulses when it is adopted in the design.

Multiple Bits, Data/Address Bus

Normally you might also want to send data/address bus across the clock domain instead of just single bit. However, things are getting more complicated when multiple bits cross the clock domain simultaneously. Simply putting n copies of two-FF synchronizers is not gonna cut it, thanks to the random nature of meta-stability. An example of this is illustrated in Figure 9-16. If two bits change from 0 to 1 on the same clock edge in clock domain 1, and assume they have caused setup violations when they are sampled by two-FF synchronizers, the outcome of each synchronizer can be either 0 or 1, since both are valid outcomes under meta-stability. Thus in clock domain 2, there is one clock cycle uncertainty between the rising edge of those two bits. In the worst case, there could be one clock cycle gap between them, as demonstrated in Figure 9-16.

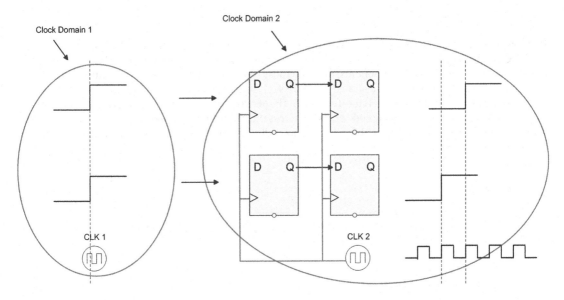

Figure 9-16. *Misaligned data, CDC of two bits*

Due to the uncertainty caused by meta-stability, you have to take extra measures to get a parallel data/address bus across the clock domain. Consider the following three approaches.

Counters with Gray Code

If the bus intended for CDC is actually a counter value, you can play tricks by using better code schemes. For a binary counter, since there is more than 1-bit flipping in its successive value sequence (e.g., when wrapping around, the counter value can go from all ones to all zeros), bit-misalignment will happen if un-coded binary values are sent directly across the clock domain. One solution is to convert binary code into *gray code* (Ref [3]), as exemplified in Table 9-2 (3-bit gray code). As you can see in Table 9-2, the benefit of gray code is that there is only one-bit difference for any two consecutive values, so the data-misalignment issue does not apply during CDC.

Table 9-2. *Three-Bit Gray Code*

Binary Code	Gray Code	Binary Code	Gray Code
000	**000**	100	**110**
001	**001**	101	**111**
010	**011**	110	**101**
011	**010**	111	**100**

Asynchronous FIFO

For a more general and reliable solution, asynchronous FIFO is often used to send data/address bus across the clock domain (Ref [3]). Although the implementation details may vary, asynchronous FIFOs are usually implemented with three basic components: block memory (dual port), read pointer, and write pointer. One way to piece them together, as demonstrated in Figure 9-17, is to let the read pointer and write pointer work on their respective clock domains as binary counters. These two pointers are used to address the dual port RAM, and they are converted to gray code and sent across the clock domain to generate FIFO status (full flag /empty flag, etc.). Thanks to the favorable traits of gray coding (as mentioned earlier), the two binary counter values can now be compared reliably across the clock domain.

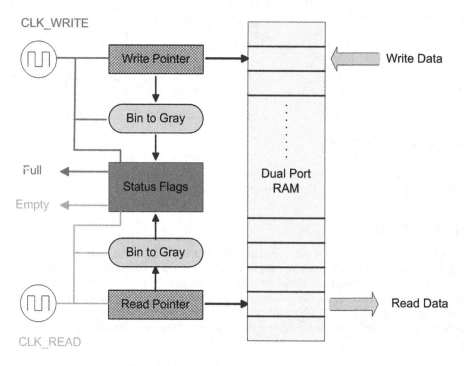

Figure 9-17. *Asynchronous FIFO*

Fortunately, it is very rare that design engineers have to get through all those tedious details mentioned here. Most FPGA design software today will offer free libraries of a wide variety of FIFOs.

Toggle Synchronizer for Data Enabled Strobes

Although asynchronous FIFO offers a full-blown solution for CDC, it also comes with overheads like block RAM and binary/gray counters. If all you need is to send a few bits across the clock domain, and if the data doesn't change very often (the minimum interval between data change is more than two clock cycles in a destination clock period), you could find yourself a lightweight solution by employing a toggle synchronizer to send data enabled strobes. The corresponding data can go straight to the destination clock domain without passing through any two-FF synchronizer, as demonstrated in Figure 9-18. If you needed to send an acknowledgement back for handshaking purposes, you could also put another toggle synchronizer in the reverse direction, as suggested in Ref [2].

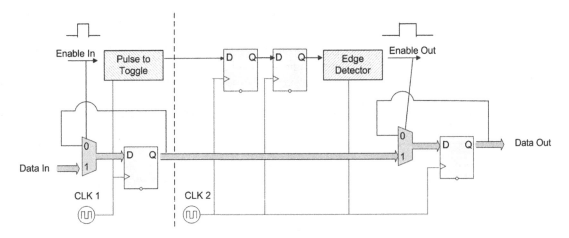

Figure 9-18. *Data enabled strobe through a toggle synchronizer*

Area-Time Tradeoff and AT² Boundary

Digital designers always face the challenge of reducing the execution time (increasing the speed) while keeping the resource usage as low as physically possible. A compromise has to be made between those two conflicting design goals.

Design for Speed

On the speed front, there are three ways to meet the design goal, discussed next.

Using a Pipeline to Increase Throughput

A pipeline can be used to increase throughput by breaking a big trunk of combinational logic into smaller ones and distributing them among multiple stages.

As illustrated in the top half of Figure 9-19, the big trunk of combinational logic will hinder the maximum clock frequency (FMax) that can be achieved. For the sake of argument, let's ignore the setup time and hold time for flip-flops (assume zero inertial delay for the flip-flops) and assume the big combinational logic has a transport delay of Δ ns. Now if you add one more pipeline stage and break the big combinational logic evenly into two smaller ones, as shown in the bottom half of Figure 9-19, the transport delay between flip-flops will become $\Delta/2$ ns. Consequently, the FMax will be doubled thanks to the pipeline. And now you can drive the pipelined version with a clock rate that is twice the one of none-pipelined version. Assume the design has a valid output on every clock cycle. The none-pipelined version has a maximum output rate of $1/\Delta$, while the pipelined version has $2/\Delta$.

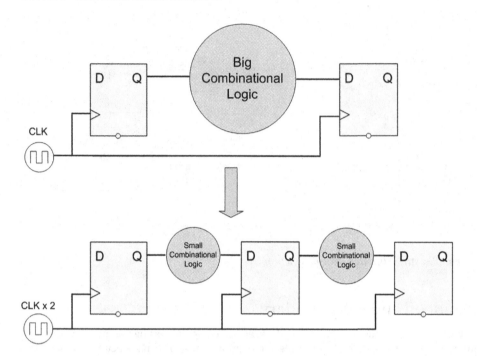

Figure 9-19. *Using a pipeline to increase throughput*

Although manual optimization can always be adopted to divide the big trunk of combinational logic into small trunks, the whole process will be tedious and error-prone. A less labor-intensive way is to use the "register retiming" in physical synthesis, as demonstrated in Figure 9-20. By inserting a few empty pipeline stages in front of the big trunk of combinational logic and turning on the register retiming, the physical synthesizer can move combinational logic freely between flip-flops to optimize the timing.

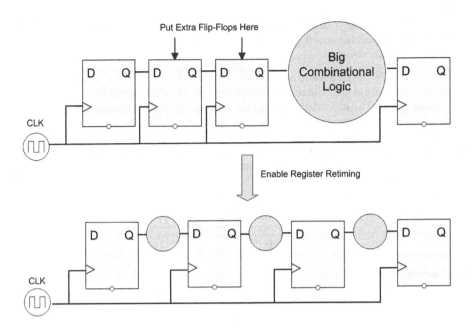

Figure 9-20. *Using register retiming for a pipeline*

Using Parallelism to Increase Throughput

Just like adding extra lanes to the freeway will ease traffic jam, adding extra processing units will surely help boosting the throughput. As shown in Figure 9-21, assume the processing unit has a processing time of Δ ns. The top half of Figure 9-21 has a throughput of $1/\Delta$. By putting two extra copies of processing units in parallel with the existing one and using arbiter and multiplexer to direct the in/out, the bottom half of Figure 9-21 can achieve a throughput of $3/\Delta$.

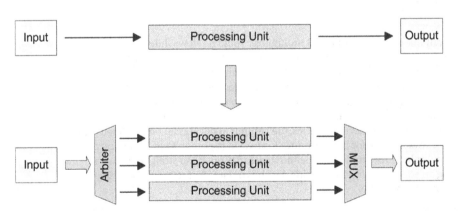

Figure 9-21. *Using parallelism to increase throughput*

221

If the data path in Figure 9-21 is wide, such as 32-bits or more, and if the number of processing units is also big, the arbiter and multiplexer will become the bottlenecks for timing closure. The following are a few useful tips for arbiter and multiplexer design:

- For the arbiter, it usually can be done in a round-robin fashion for fairness. As shown in Figure 9-22, a one-hot circular shift register can be used for the input enable of each processing unit, and an input data bus can be shared among all the processing units. If the number of processing units is big and a large fan-out is needed, designers can take advantage of the "register duplication" feature in physical synthesis by inserting a few empty pipeline stages on the data bus, and the physical synthesizer should be able to take care of the rest and duplicate registers when a large fan out is desired.

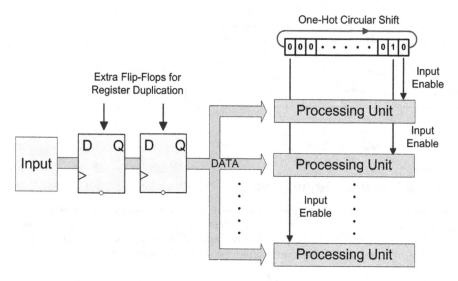

Figure 9-22. *Using parallelism to increase throughput*

- For the multiplexer, a traditional N:1 multiplexer could be slow while occupying a large area. Instead of using the traditional approach, an alternative solution is to use XOR-adder tree. As illustrated in Figure 9-23, the same one-hot circular shift register from Figure 9-22 can be used as the output enable for each processing unit. Those 2:1 multiplexers shown in Figure 9-23 can be replaced by an AND gate if necessary. The outputs of the 2:1 multiplexers are summed by an N-input XOR gate. Either manually or through physical synthesis, the N-input XOR gate can be pipelined into a multi-stage adder tree for better timing closure. On the other hand, chopping a big N:1 multiplexer into multiple pipeline stages may not be so straightforward as chopping an adder tree. Hence for wide data bus or large N, I recommend using the XOR-adder tree structure.

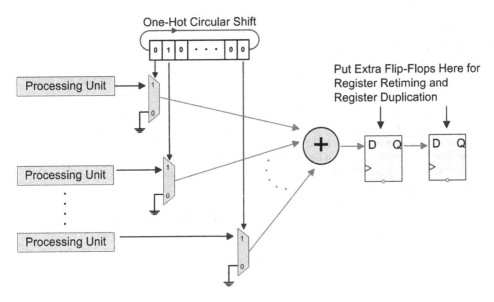

Figure 9-23. *Using XOR adder tree to replace the N:1 multiplexer*

Using a Ping-Pong Buffer

A ping-pong buffer is a special case of Figure 9-22, whereby *N* equals 2 and the processing units are just simple-dual-port block RAMs. Under such circumstances, one block RAM is called the *ping* buffer and the other is called the *pong* buffer. At any given moment, one and only one of them is taking the input (data write), while the other is doing the output (data read). Those two block RAMs don't have to be physically separated. Instead they can be combined into a bigger circular buffer, with the read pointer trailing the write pointer by half the buffer size, as illustrated in Figure 9-24.

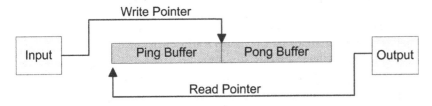

Figure 9-24. *Ping-pong buffer*

Design for Area

If you have a brimming FPGA, a first-aid solution is to convert some most scarce resources into other types, such as turning block RAMs into logic or vice versa, or converting DSP blocks into logics. If you can't find a quick way out after that, architecture overhaul has to follow. And it usually means resource reuse and doing things in multiple cycles.

- Turning the pipeline to a multiple-cycle implementation. Unfortunately, as far as I know, this has to be carried out on an ad hoc basis. A typical example can be found in the FIR filter implementation. If the design goal is the speed, a transposed architecture like the one shown in Figure 9-25 is often adopted. Assume N = 64 and assume both input data and coefficients are 16-bit 2s complementary. For the sake of simplicity, no guard bits are used and overflow is not considered, thus 32-bits will suffice for each register in Figure 9-25. The total number of flip-flops hence becomes $64 * 32 = 2048$, and total number of multipliers is 64. A throughput of one data output per clock cycle can be achieved with such a transposed architecture.

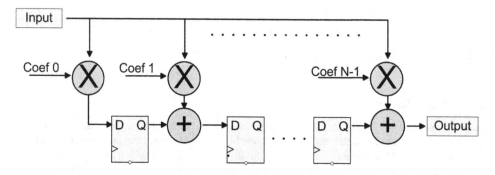

Figure 9-25. *FIR filter—transposed architecture*

- On the other hand, if resource usage is more of a concern, a much smaller design can be realized with just one MAC (multiplier-accumulator), as suggested in Figure 9-26. If you don't take overflow into account, one 32-bit wide register will suffice. The total number of multipliers is just one. But the smaller design is also much slower. Its throughput is only 1/64 of that of transposed architecture's.

- Using CPU core to replace logic. A more generic solution to save space might be to use CPU core instead of logic. This approach can be adopted for large-scale signal processing where the control is irregular and convoluted. Under such circumstances, the initial resource cost for building the CPU core can be offset by the high rate of ALU reuse. Ref [18] provides a good example to implement vocoders in FPGA using DSP cores.

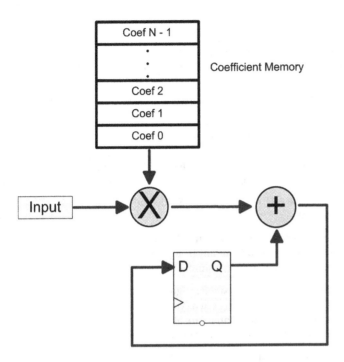

Figure 9-26. *FIR filter—MAC architecture*

Area-Time Tradeoff and AT² Boundary

From previous sections, the intuitive conclusion is that you can have a design that runs super-fast, or a design that is extremely small in size, but at the end of the day, you have to make a compromise somewhere between the super speed and the small area. So this begs the question: Can you come up with a design that has the best of both worlds?

The answer is negative, as you might have expected. In fact, people have already tried to find answers for this question in the semiconductor industry's early days. Those pioneers in Ref [16][17] had come to the conclusion that:

For a given semiconductor process, if the design has an area of A and an execution time of T, there exists a lower boundary B that $AT^2 \geq B$.

So like other engineering works, it is all about give and take. You can't have your cake and eat it too.

Inter-FPGA Communication

If you have a large design that has to be partitioned into multiple FPGAs, and if you need to move data between FPGAs in high throughput, this section might provide some useful clues for you.

Parallel Source Synchronous Bus

If you have enough pins available on both FPGAs, you can use parallel source synchronous bus for data transfer. But make sure you have done enough board-level simulations before you commit to this. For all the caveats on corresponding FPGA setup, refer to the previous chapters that discussed "Bus with Source Synchronous Timing".

Using Serial Links

On the other hand, differential serial links can be used to provide a data path for both TX and RX. In this regard, if your FPGA has a hard-core SERDES on board, take advantage of it by all means. Otherwise, a low-cost alternative is to use a regular differential transceiver (like the one in Ref [19]) plus some bit alignment circuit to form the link, as described in the following[6]:

1. Choose differential pairs dedicated to TX and RX.

2. The clock input for both FPGAs should come from the same oscillator.

3. Register the input and output in the IO pad. (For Altera FPGA, it means enabling fast input/output registers in the assignment setting.)

4. It is better to enable DPA (Dynamic Phase Alignment) on the RX side and wait until all the channels are locked on DPA.

5. After DPA is locked, the word boundary should be aligned. In other words, the bit-slip should be handled for each channel individually by toggling the rx_channel_data_align pin or other control pins with a similar function.

    To handle bit-slip, the TX should be composed of two states: alignment and normal. In alignment state, the TX should keep sending out a sync word, such as 0x5C until it gets an indication from RX that all channels are aligned. Such an indication can be realized by using a GPIO pin from RX back to TX.

6. The RX should keep doing the word boundary alignment until:

    a. All channels are receiving the sync word correctly

    b. The sync word has been successfully received on all channels for a continuous number of cycles (set a threshold for this)

7. After the alignment is finished, the RX should notify the TX through something like a GPIO so that the TX can start its normal transmission.

8. Both TX and RX can be set with 180-degree phase offset between data and clock.

9. On the RX side, the data is received on the TX clock domain. An asynchronous FIFO can be used to transfer the data to another clock domain for further processing. And start reading data out of the FIFO when the FIFO is half full.

Protect Your IP

For any company in the world, intellectual property and trade secrets are always closely guarded for obvious reasons. The software industry has been plagued by piracy since day one, and schematic/PCB designers fare no better as their works are often artfully reverse-engineered (Ref [20]). As for FPGA designers, if you are selling your modules separately as standalone IPs, two things should be heavily protected from your rivals and imitators: the simulation library and the net list.

---

[6]The solution described here mainly applies to Altera FPGAs. But I guess similar setups can be found for devices from other FPGA vendors.

Encrypt Your Source Code

Most EDA vendors today will offer options to encrypt your VHDL/Verilog code. Usually it is done through public/private key pairs. If you are using Modelsim for simulation, you can encrypt your code by enclosing them in encryption envelopes (Ref [21]), and then you can compile the code with the +protect option. For delivering simulation library, encryption is a good choice. Refer to the EDA vendor's user manual for more options on source code encryption.

Obfuscate Your Source Code

In addition to the simulation library, your net list could be another source to divulge your secret, since the names used in the net list are often derived from the names of your design entity or modules. And the net list could also reveal your design hierarchies if no proper measures are taken.

To confuse the potential crackers, you could choose to obfuscate your source code. There are professional tools that are capable of doing this (Ref [22][23]). And while you are obfuscating your source code, don't forget to do that to the filenames as well.

To hide your design hierarchies, make sure you flatten your design during synthesis/compile stage. For Altera Prime, the hierarchy will automatically be flattened if you don't create any design partitions (Ref [24]). And if you are using Synplify, you can set the syn_hier attribute to flatten (Ref [25]). For other tools, refer to the vendor's document for more information.

On top of that, it is also possible for the crackers to get some clues by watching the internal signal of your simulation. To make their job harder, the internal simulation models should be hidden. If you are using Modelsim, the -nodebug option can be used to hide internal model data so that *all source text, identifiers, and line number information are stripped out* (Ref [21]).

Protect Your Bit Stream

IP thieves may also go after your bit stream files so that they can make an illegal duplication of your product. And the possibility is heightened when manufacturing is not done in-house. To protect your bit stream files, you can do the following.

Encrypt Your Bit Stream

Some FPGA devices allow users to store the encryption key inside FPGA. The encryption key can be saved as one-time programmable fuse, or it can be stored in non-volatile memory that is write-only. The encryption/decryption algorithm adopted for those cases is usually AES.

As for the place to store the bit stream, it can be Flash memory or a MCU that resides outside the FPGA, as shown in Figure 9-27. Or the bit stream can be saved inside FPGA if the FPGA is Flash-based, like Altera MAX 10 device.

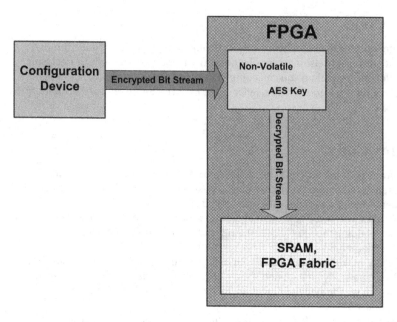

Figure 9-27. *Bit stream encryption and decryption*

Node Lock Your Bit Stream

Some FPGA devices, like Altera MAX 10, contain a unique chip ID for each device. You can make your design functional only with certain chip IDs. This approach is very similar to the way software companies node-lock their licenses to certain MAC addresses.

If you choose to use the node-lock approach, you might have to build a customized bit stream file for each device. Of course, you can always do that through logic design (VHDL/Verilog) and go through the synthesis/PAR/Assemble every time. However, if you have a huge project to build, the whole process would be time consuming, which may not be feasible for large-volume production.

Instead, you could put a soft core processor inside, and let the processor verify the chip ID. In this way, you only need to replace the memory init file each time. Thus the two most arduous steps (synthesis and PAR) could be skipped completely. And as suggested in previous chapters, don't forget to check the CRC checksum in this case, since someone could tamper with your memory content to bypass the ID verification.

Limit JTAG Access

JTAG could be another source of vulnerability. Crackers could snoop around through JTAG, or program the FPGA with a tampered version of their own. So limit the JTAG access if your FPGA devices support that.

Anti-Tamper

If for some reason the FPGA device you choose lacks the desired security features, you could protect your IPs at the system level. Because they are done at a higher level, the measures discussed in this section can be applied universally to all embedded systems.

Anti-Tamper Sensor/Switch

As illustrated in Figure 9-28, the whole system can be kept in a sealed environment, such as a case. And an anti-tamper switch or sensor, like the one in Ref [29], can be installed beneath the case cover. Whenever the cover is removed, the anti-tamper sensor/switch will be triggered, and the MCU will be notified.

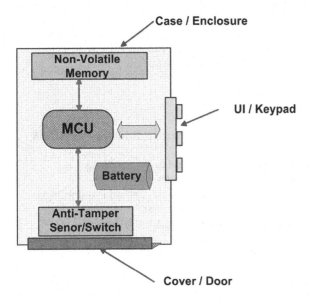

Figure 9-28. *Anti-tamper sensor/switch*

If the cover/door is removed or cracked open without proper authorization (keying in the passcode, for example), the anti-tamper switch will drive MCU to take counter-measures, such as zero-out all the sensitive data on the flash and put the system into a non-functional state. For FPGA, it usually means the bit stream files will be erased from the configuration device, or the AES keys will be deleted.

Bury the Conductive Wire

By observing Figure 9-28, you will notice that the keypad is not fully sealed. If you are as paranoid as I am, you might want to protect it from snooping as well. One thing you can do is to bury those conductive traces (under the key plungers) into the inner layers of your PCB, and your keypad could use capacitive sensors instead of conductive ones, as suggested in Ref [30]. By doing this, you can make your passcode input less prone to keypad snooping.

Of course, copy cats these days are more resourceful than what I can imagine. But by taking these measures, their progress can be hindered a great extent.

Summary

The main topics of this chapter were FPGA and IP protection.

There are two main HDLs (hardware description languages) that can be used to program FPGA. This book is leaning toward Verilog/Systemveriolg. To get good performance in digital design, the equation of Performance = Parallelism + Pipeline is a good one to keep in mind. A good design is always a compromise between speed and area.

To protect your FPGA design, you can take measures to encrypt or obfuscate your netlist and remove useful information from simulation library, or you can use node-lock to guard your bit streams. In a more general sense, you can take anti-tamper measures to thwart crackers and copycats.

References

1. "Crossing the abyss: asynchronous signals in a synchronous world." Mike Stein, Paradigm Works, *EDN Magazine,* July, 24, 2003
2. "Practical design for transferring signals between clock domains." Michael Crews and Yong Yuenyongsgool, Philips Semiconductors, *EDN Magazine,* February, 20, 2003
3. Simulation and Synthesis Techniques for Asynchronous FIFO Design, Rev 1.2, Clifford E. Cummings, Sunburst Design, Inc., SNUG (Synopsys User Group Conference), San Jose, 2002
4. *Verilog Digital Computer Design - Algorithms into Hardware.* Mark Gordon Arnold, University of Wyoming, Prentice Hall PTR, 1999
5. *Verilog HDL Synthesis - A Practical Primer.* J. Bhasker, Bell Labs, Lucent Technologies, Star Galaxy Publishing, 1998
6. Verilog Digital System Design, Zainalabedin Navabi, Northeastern University, University of Tehran, McGraw-Hill, 1999
7. *The Designer's Guide to VHDL, Third Edition.,* Peter J. Ashenden, Morgan Kaufmamn Publishers Inc., May 2008
8. *Designing with FPGAs & CPLDs.* Bob Zeidman, CMP Book, 2002
9. *Verilog Designer's Library.* Bob Zeidman, Prentice Hall PTR, 1999
10. *SystemVerilog for Design Second Edition: A Guide to Using SystemVerilog for Hardware Design and Modeling (Second Edition).* Stuart Sutherland, Simon Davidmann, and Peter Flake, Springer Science+Business Media, October, 2010
11. *SystemVerilog for Verification: A Guide to Learning the Testbench Language Features (Second Edition).* Chris Spear, Springer Science+Business Media, November, 2010
12. "Microcontroller Oscillator Circuit Design Considerations." Cathy Cox and Clay Merritt, Freescale Semiconductor, 2004
13. TCXO Application vs OCXO Application (Application Notes #803, REV 1), Dave Kenny, PLETRONICS
14. "Synchronous Resets? Asynchronous Resets? I am so confused! How will I ever know which to use?". Clifford E. Cummings and Don Mills, SNUG (Synopsys Users Group), San Jose, 2002
15. "Asynchronous & Synchronous Reset Design Techniques - Part Deux." Clifford E. Cummings, Don Mills, Steve Golson, SNUG Boston 2003
16. Area-Time Complexity for VLSI, C. D. Thompson, Carnegie-Mellon University, CALTECH CONFERENCE ON VLSI, January, 1979
17. Some Area-Time Tradeoffs for VLSI, Richard P. Brent and Leslie M. Goldschlager, SIAM (Society for Industrial and Applied Mathematics) *J. Comput.,* Vol. 11, No. 4, November, 1982
18. FPGA implementation of a DSP Core for Full Rate and Half Rate GSM Vocoders, Hamid Noori, Hossein Pedram, Ahmad Akbari, Shervin Sheidaei, The 12th International Conference on Microelectronics, Tehran, October 31 - November 2, 2000
19. LVDS SERDES Transmitter/Receiver (ALTLVDS_TX and ALTLVDS_RX) IP Cores User Guide, Altera Corp, December, 2014

20. The Art of PCB Reverse Engineering: Unraveling the Beauty of the Original Design, by Mr. Keng Tiong Ng, CreateSpace Independent Publishing Platform, February, 2015

21. ModelSim User's Manual, Software Version 10.1c, Mentor Graphics Corporation, 2012

22. Thicket(TM) Family of Source Code Obfuscators, Sematic Designs, Incorporated (http://www.semdesigns.com)

23. HDL Code Obfuscation (Tcl Script), Aldec Inc. (https://www.aldec.com)

24. Quartus Prime Standard Edition Handbook Volume 1: Design and Synthesis, Altera Corporation, May 4, 2015

25. Synopsys FPGA Synthesis Synplify Pro for Microsemi Edition - User Guide, Synopsys, Inc. February, 2013

26. AN 593: Anti-Tamper Protection for Cyclone III LS Devices, Altera Corporation, October, 2009

27. MAX 10 FPGA Configuration User Guide, Altera Corporation, December 14, 2015

28. White Paper: Anti-Tamper Capabilities in FPGA Designs, Ver 1.0, Altera Corporation July, 2008

29. Omron D2FS Ultra Subminiature Anti-Tamper Switch, OMRON Corporation, 2013

30. AN528: Buried Capacitive Sensors for Tamper Protection, Rev 0.2, Silicon Laboratories Inc., July, 2013

## CHAPTER 10

System on Programmable Chip (SOPC)

Where a calculator like ENIAC today is equipped with 18,000 vacuum tubes and weighs 30 tons, computers in the future may have only 1,000 vacuum tubes and perhaps weigh only 1½ tons.

—*Popular Mechanics* magazine, March 1949

Programmable hardware, as the book's title suggests, starts to emerge as an important part of embedded system design. And more specifically, the SOPC (System on Programmable Chips) approach is moving toward the mainstream, thanks to decades of groundwork laid by the FPGA industry. Given all that, this chapter explains why SOPC makes more sense for your next big design.

Economics 101

For those of us who are not immortal, making a viable product in a profitable way is the principle we follow on a daily basis. Thus we always keep a close eye on products' BOM (Bill of Materials) costs. In the old days, FPGA devices were often conspicuous by their absence in the BOM, mainly because:

- Device density was low at that time. Those devices did not have enough logic or memory resources to hold a soft core microcontroller inside.

- Those early devices were mostly SRAM-based. So a separate device, such as a Flash memory, has to be present to store all the configuration data. This standalone configuration device will not only increase the BOM cost, it will also open doors to unauthorized copying.

- There were no analog components, such as ADC, integrated in those devices.

- Compared to COTS (Commercial Off The Shelf)[1] components, the price of FPGA devices was extremely high at that time.

---

[1]Generally speaking, FPGA can also be counted as COTS. But in this book, COTS is used to refer to ASIC and general-purpose processors.

© Changyi Gu 2016
C. Gu, *Building Embedded Systems*, DOI 10.1007/978-1-4842-1919-5_10

However, things have changed a lot since then. To show that SOPC approach makes economic sense now when it comes to sophisticated systems, I will do a side-by-side cost analysis between the COTS and SOPC systems. On the COTS side, I use the popular Arduino UNO Rev3 (Ref [1]) as its representative, while on the SOPC side I use the Altera MAX 10 device (Flash-based FPGA) as the base of implementation.

As many of you know, the component's unit price and its corresponding volume price (1k price, for example) could differ a lot, thanks to the economies of scale. To compare apples to apples, all the prices I quote in this chapter are unit prices and publicly available from the vendor or distributor web sites. Keep in mind that component prices are not set in stone. By the time you are reading the published version of this book, these numbers could have deviated from the reality, as inflation can go both ways, and the economy waxes and wanes. Since I'm not the chairman of the Federal Reserve, don't hold me accountable if those numbers do not reflect the latest economic truth. They are listed here only to give you an idea in a relative sense.

MCU vs. Soft MCU

The Arduino UNO Rev 3 employs an ATmega328P MCU from ATMEL. From the Digi-Key web site, its unit price is $3.38. (Its 2k price is $1.675 in case you are curious.) This 8-bit MCU has a plethora of analog and digital peripherals, with maximum clock rate of 20MHz. On the other hand, a soft MCU core can be implemented in FPGA. For an Altera device, you can choose a NIOS II processor (32 bit), with less than 700LEs. Or you can choose a 8051 core (8-bit, four clock cycles per instruction) with less than 2000LE. Thus an Altera 10M02 device can be used with a unit price of $3.78. The side-by-side comparison is shown in Table 10-1.

Table 10-1. *Cost Analysis for COTS MCU vs. Soft MCU*

Item	Arduino UNO Rev 3	Altera MAX 10
Main component	*ATmega328P ($3.38)*	*10M02-C8 ($3.78)*
Processor core	*AVR @ 20MHz*	*NIOS II (Economy) or 8051 (4T) @ 100MHz*
ADC	*1*	*0*
Total price	**$3.38**	**$3.78**
Winner of price war	X	

Clearly, if MCU is the only major component of your system, using COTS (a general purpose processor) makes more economic sense.

MCU Plus External Peripherals

General-purpose MCUs today contain varieties of peripherals that can cover most common applications. However, one size cannot fit all, which is always the soft belly for general-purpose MCUs. For the example at hand, Arduino UNO needs to communicate with the host through a USB port (USB-based UART), for which it uses ATmega16U2 (Ref [2]).

The ATmega16U2 is another AVR microcontroller with USB 2.0 full-speed module built-in. Its unit price is $4.35. (For USB UART, there are other alternatives as mentioned in previous chapters. One of them is offered in Ref [3] with a unit price of $4.50.)

The counter-offer from SOPC/FPGA side is to implement a USB MAC layer inside FPGA, with an external USB PHY transceiver. A typical USB MAC might cost a little more than 2000LEs, so the Altera 10M04 device can be used to host both the soft MCU and USB MAC, with the assistance of an external ULPI USB transceiver. The unit price for 10M04 devices (no built-in analog component) is $11.43, while an ULPI USB transceiver like the one in Ref [4] costs $1.24 per unit.

Now assume you need to do a little more, such as send data to other modules through an LVDS interface (parallel to serial). For FPGA, LVDS is a cinch for its IO capability, and the parallel to serial function can be done with less than a few hundred LEs. However, for the AVR microcontrollers used by Arduino, an external shield with LVDS transceiver has to be mounted. For the sake of simplicity, let's skip the PCB cost of the shield and only take its IC cost into account, which could be the one from Ref [5] with a unit price of $6.79 (transmitter only).

The total tally is summarized in Table 10-2. With external peripherals joining the price war, the scale is quickly tipping toward the FPGA side. As I said earlier, the soft belly of general-purpose MCU is that its peripheral portfolio cannot be customized. Once you have difficulty finding a MCU that can cover all the peripheral needs, the external components added will bump up the BOM cost in big ways. On the other hand, SOPC/FPGA can always be molded to target a specific application.

Table 10-2. *Cost Analysis for MCU with External Peripherals*

Item	Arduino UNO Rev 3	Altera MAX 10
Main component	*ATmega328P ($3.38)*	*10M04-C8 ($11.43)*
Processor core	*AVR @ 20MHz*	*NIOS II (Economy) or 8051 (4T) @ 100MHz*
USB	*ATmega16U ($4.35)*	*Soft MAC, External PHY - USB3300 ($1.24)*
LVDS transmitter	*SN65LV1023A ($6.79)*	*Soft Parallel to Serial*
Total price	**$14.92**	$12.67
Winner of price war		X

Multiple Processors

Now let's seek a more sophisticated application, like an MP3 player. For this case, two processors are needed. As always, one of them will be a MCU, which handles the interface and house-keeping matters. The other can be a standalone DSP processor or it can be a SOC/ASIC with a DSP core inside. For Arduino, it means adding an MP3 shield, like the one offered in Ref [6], with a unit price of $35. Its core component is actually an SOC/ASIC that has a DSP processor core and an analog codec integrated together (Ref [7]). From the vendor web site, this chip sells with a unit price of $20 ($10 at 100 pieces, and a 1k price of $3.10).

To match that with SOPC/FPGA, you can use a bigger device like Altera 10M08, adding a NIOS II Fast core with DSP accelerators. However, the analog codec chip has to be done with external components, like the one offered in Ref [8]. The 10M08 device (C8 speed grade) has a unit price of $14.88, while the codec chip is $2.82.

Again, let's skip the PCB cost of the Arduino and only take IC cost into account. The final tally is summarized in Table 10-3.

Table 10-3. *Cost Analysis for MP3 Player Application*

Item	Arduino UNO Rev 3	Altera MAX 10
Main component	ATmega328P ($3.38)	10M08-C8 ($14.88)
MCU core	AVR @ 20MHz	NIOS II (Economy) or 8051 (4T) @ 100MHz
USB	ATmega16U ($4.35)	Soft MAC, External PHY - USB3300 ($1.24)
MP3 solution	SOC, VS1053 ($20.00)	NIOS II (Fast) External Codec - MAX9867 ($2.82)
Total price	**$28.13**	**$18.94**
Winner of price war		X

At first blush, it looks like the SOPC/FPGA is holding a commanding lead over SOC/ASIC solution in cost, which runs counterintuitive. All right, I have to confess that I've played a few accounting tricks here. As I stated earlier, the prices you see in Table 10-3 are all quoted as unit prices. However, if you take a close look at the price curve of Ref [7] from the vendor web site, its price will drop to $10 if you buy more than 100 pieces, $3.10 at 1K volume, and $2.74 at 9K. Of course, the SOPC/FPGA device will also have a falling price when the volume goes up. The Altera 10M08-C8 may drop to $8.70 if you buy hundreds, but I don't think its price curve would be as steep as that of SOC/ASIC. In other words, the SOPC/FPGA may still hold its line if the volume is on the order of hundreds. But for anything beyond that, the SOPC/FPGA will lose the price war.

Here is my take on this: The MP3 player is a mainstream application, for which plenty of SOC/ASIC solutions can be found. And SOPC/FPGA cannot compete with SOC/ASIC on cost once the volume is large. However, if your application is a little bit off the mainstream, where no SOC/ASIC exists, the SOPC/FPGA can be a good choice in terms of cost, especially when the application is sophisticated and needs some level of customization.

Of course, if your application later starts to be accepted by the mainstream and the volume goes really high, the SOPC/FPGA solution can be easily turned into SOC/ASIC. But given the high NRE cost today, the breakeven point could be on the order of 10^5 - 10^6. Consult your end-to-end semiconductor solution company for a detailed cost analysis.

Processor Core

For anything to be called an embedded system today, it must contain some form of processor core inside. And some systems could even carry more than one processor along with a wide variety of peripherals. This section discusses the available processor choices you have for SOPC. The next section will discuss the bus architecture.

For the processors to be used in SOPC, the current available choices are as following:

- *Use the hard-core processor provided by the FPGA vendors.*

  FPGA vendors today are providing devices that have hard-core processors inside, such as Altera's HPS SoC or Xilinx's Zynq SoC. Those high-performance processors are usually ARM cores, with a clock rate above 1GHz. If you can afford the price tag, go for it by all means.

- *Use the soft-core processor provided by the FPGA vendors.*

  FPGA vendors today also provide soft-core processors that can be used as IPs in their devices, such as Altera's NIOS II or Xilinx's MicroBlaze. Those cores can be put into vendors' low-end devices as long as they fit, and they are very flexible in terms of sizing and performance. They can often be driven with a clock rate of 100MHz or higher.

  The plus side of those vendor-provided soft-cores is that their licensing terms are often generous, and they come in handy for low-to-mid-range applications. But their drawbacks are as follows:

  1. Their ecosystems are not as big as that of ARMs or 8051s.

  2. They are not portable, and you are tied to specific vendor once you get on board with them.

  3. If by any chance you want to make an ASIC out of current FPGA design, you might have to negotiate a separate licensing terms for the processor core to be used in ASIC.

- *Jump onto the ARM bandwagon.*

  It is always possible to put an ARM soft core inside FPGA. In fact, the first of such cores I'm aware of is ARM Cortex M1 (Ref [9]). However, for some reasons unknown to me, I had a hard time gathering publicly available information for this particular ARM core. At this point, my best guess is that in its smallest configuration, it has an area and clock rate that is comparable to a 1T 8051 core[2].

  Generally speaking, ARM core is not a bad choice thanks to its popularity in cell phones and tablet devices. But its popularity comes with a caveat: You have to negotiate a good term with ARM for license fee and royalty, which could be a sum in the order of millions. (Again, talk to ARM directly for the most up-to-date number, and don't hold me accountable for any wrong predictions.) And go for it as long as you think it makes economic sense for you.

- *Go with the good old 8051.*

  Throughout the history of human beings, wise men have predicted the end of the world on more than one occasion. So is the demise of 8-bit MCUs. Among those surviving breeds, 8051 stands out as a good, low-cost choice for SOPC/FPGA implementation. I will have another section dedicated to 8051 in later part of this chapter. For now, remember the following:

  1. Unlike ARM, you don't have to sign a license/royalty agreement before you implement the 8051 instruction set. And if you later decide to upgrade your SOPC/FPGA to SOC/ASIC, you will not be burdened by a license fee/royalty either.

  2. 8051 is not tied to any specific FPGA vendor, so it is very portable.

  3. 8051 has a very big ecosystem. There are more than 70 vendors (Ref [10]) worldwide that make various enhanced versions of 8051. Most of them are doing RISC implementation of a CISC instruction set, just like Intel has done to its x86 processor. So the 8051 you are using today is not the same as that of your grandma's.

- *Other possibilities, such as LEON.*

  There are always other possibilities beyond those mentioned. LEON (Ref [16]) is one of them. It is a 32-bit, open source processor with a SPARC V8 architecture. However, note that open source does not necessarily mean free. You might still have to seek a license agreement before you put it into commercial use.

---

[2]1T 8051 core means most instructions (barring branch instructions) are completed in one clock cycle.

Interconnection

To glue peripherals around the processor, you will also need to choose a bus architecture. Your choices are more limited in this case:

1. *AMBA (Advanced Microcontroller Bus Architecture)*

   If you've decided to use ARM or MicroBlaze as your processor, AMBA is pretty much your only choice. AMBA includes multiple specifications to cover devices of various kinds (Ref [11][12][13][14]). Fortunately, most FPGA vendors will have tools to help you wrap your device/IP with the targeted AMBA specification.

2. *Avalon*

   If you've decided to use NIOS II as your processor, Avalon (Ref [15]) is also your only choice for bus architecture. Altera has provided QSys tool to stick multiple components together through Avalon.

3. *Wishbone*

   The Wishbone bus (Ref [17]) was developed as an open source bus architecture for SoC interconnection. And it is supposed to be less burdened by patents.

Design and Use Your Own 1T 8051

As I mentioned earlier, 8051 can be a good choice for low-end SOPC/FPGA design. Unlike ARM processors, you don't have to obtain any license permissions before you start designing your own 8051 cores. In fact, there have been more than 70 vendors in this world who are offering (or once offered) 8051 in clones or various enhanced flavors, including folks from the once almighty USSR. This section discusses how to design a 1T 8051 and how to incorporate it into your own system.

History 101

The first 8051 was born in 1980 at Intel, as a successor to the 8048 microcontroller. And it has gained its popularity since then. Consequently, it has been licensed to other manufacturers worldwide (Ref [19]). The original 8051 was produced using the NMOS process, with later versions starting to adopt the CMOS process. The original 8051 stores its programs in external ROM, and nowadays Flash-based 8051 MCUs have become universal.

The 8051 has a Harvard Architecture and a CISC instruction set. In its original design, it ran at a clock rate of 12MHz (or perhaps 11.0592 MHz), and it took 12 clock cycles to execute each instruction. (I guess at that time there was no such concept of RISC or pipeline, but I could be wrong.) Thus the original 8051 is often called 12T 8051. As you can see, it is extremely slow and inefficient by today's standards.

With the advance in CMOS technology, new applications are crying for more efficient implementation with higher clock rate. Thus the enhanced models start to mushroom. And in later sections, I will discuss how to design an enhanced 8051 using RISC implementation. In other words, I'm going to use pipeline to boost both the efficiency and clock rate. The design goal is to complete most instructions (barring the branch, mul/div, and BCD instructions) in one clock cycle, with a targeted clock rate of 100MHz on Altera MAX10 C-8 grade devices. Also, only block RAM internal to FPGA will be used as program/data memory.

Patents and Copyrights

IP (intellectual property) rights of other companies should be respected. Before you start your own processor or microcontroller implementation, you'd better make sure you are not going to violate any IP rights. On this particular matter, I have done some legal research and I would like to share my notes here. However, keep in mind that I'm not an intellectual property lawyer. Anything I have mentioned regarding intellectual property law should be taken only on reference basis. Contact your company's legal department for professional advice or suggestions.

With that being said, there are basically three weapons that companies can take to defend their IPs: patents, copyrights, and trademarks. Trademarks are beyond the scope of this book, and I will leave it to inquisitive readers and their lawyers. As for the patents and copyrights, the patents is used to protect an idea, while the copyrights is used to protect the expression or manifestation of an idea, but not the idea itself.

SIDE NOTES

Personally I think it is a very genius design at the system level to distinguish between an idea and the expression of an idea. Legally, this is called "idea-expression dichotomy". In this regard, another genius system design I have seen so far is the "separation of powers".

One major difference between patents and copyrights is that the protections they offer have different time limits (durations). In the United States, patents are protected for 20 years, starting from the filing date. And the patent will fall into public domain after it expires.

On the other hand, the copyright protection may have various terms depending on when the work was created and the nature of the authorship. In these circumstances (assuming the technology you are interested in came into being after 01/01/1978), the shortest protection term will be life plus 70 years[3] (Ref [21]). That is the reason why I still cannot find a legitimate clone for my all-time favorite video game: Super Mario Bros.

SIDE NOTES

Although NES game console (Nintendo Entertainment System) that first carried Super Mario Bros has been discontinued for more than 10 years, the game's original music and art drawing are still under copyright protection. And on this matter, Nintendo does not hesitate to stamp out illegal clones through his army of lawyers, even for a game made in the last century (Ref [22]).

So regretfully, my chance to see a legitimate clone of Super Mario is pretty low, unless Congress passes new legislations during my lifetime.

For a processor to be useful, you need at least two things: the ISA (Instruction Set Architecture) and a physically available processor that can understand and execute the ISA. When patents and copyrights are applied to them, the deal is as following:

- As far as I know, ISA is not copyrightable. But ISA can be protected by patents.

- However, the form of processor implementation, such as VHDL/Verilog code, lay out, die mask, etc., are protected by copyright.

---

[3]That means 70 years after the death of its original author.

Since 8051 was born in 1980s, its ISA patents (if there are any) should have all expired by now. The original 8051 was a 12T processor produced through NMOS process, while the design we are trying to achieve is mainly for FPGA with 1T efficiency in mind. Thus the chance of a copyright lawsuit from Intel is next to zero, although Intel has a long history of suing everyone for everything (Ref [23]).

However, as I mentioned, there are a plethora of vendors who offer enhanced 8051. And those enhanced versions could carry an outstanding patent. But given the long history of 8051 and the plenitude of open source 8051 cores, the risk of stepping on a standing 8051 patent is pretty low. Yet again, I'm not a patent lawyer, so do a thorough patent search if you have doubts.

Having said that, I would say it is a safe bet to go ahead and start this 1T 8051 design, without worrying too much about potential patent/copyright lawsuit.

BTW, ARM clones are also blessed and troubled by patent/copyright laws. Currently, there is an open source clone of ARM v2a ISA, which is called Amber (Ref [24][25]). It seems ARM has tolerated its existence so far, given the fact that ARM v2a ISA is such an old architecture (more than 20 years old). However, other attempts to clone more update-to-date ARM architectures were less fortunate in this regard (Ref [26]).

Address Spaces

As stated earlier, 8051 has a Harvard architecture, which means its program memory and data memory are separated. The original 8051 allows a mixture of external memory and internal memory. However, per the design goal, RAM/ROM external to FPGA will not be used. So the maximum code space will be 64KB, limited by the bit width of the PC (Program Counter).

As for the data space, in the original 8051, the data spaces are divided into three separate sub-spaces: XDATA, IDATA, and SFR. The XDATA can be indexed by a DPTR register (16 bits), so it has a space limit of 64KB. The IDATA and SFR are indexed by general registers (8 bits) or immediate addressing (8 bits), and SFR (Special Function Registers) are overlapping with the upper half part of the IDATA. Thus SFR has 128 bytes of space limit, while IDATA has 256 bytes. SFR can only be accessed through direct addressing mode, while the upper half of IDATA can only be accessed through indirect addressing mode.

Now when it comes to FPGA implementation, the typical block RAM type in Altera MAX10 FPGA is M9K, which is 9 * 1024 bits ≈ 1KB. To use the block RAM more efficiently, you could merge IDATA and XDATA into one space and reduce XDATA's space limit by 256 bytes. The SFR (Special Function Registers), on the other hand, have to be implemented in logic as the name suggests.

Thus the address space for the 1T 8051 will look like the diagram in Figure 10-1.

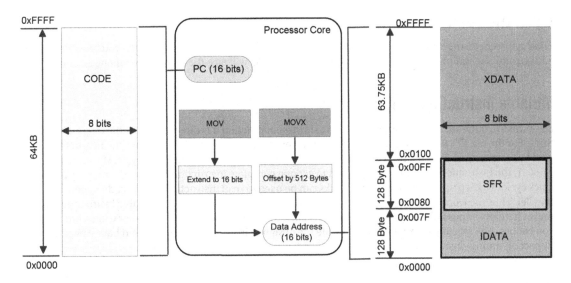

Figure 10-1. *An 8051 address space*

As illustrated in Figure 10-1, to make efficient use of an FPGA block RAM, the IDATA and XDATA are merged into the same address space and indexed by one 16-bit data address register. Thus for MOVX instruction (XDATA access), its nominal address will be shifted by 512 bytes to get the physical address. And the 8-bit address from MOV instruction (IDATA/SFR access) will be extended to 16-bit for the sake of consistency.

However, I have to point out that classic 8051 architecture has a limited stack size. In classic 8051, the stack pointer is only 8 bits. Taking into account the overhead of general-purpose registers, the stack size's theoretic upper bound is around 479 bytes (Ref [27]). However, the fact that SFR and IDATA partially overlap has made the situation even more complicated. Due to the partial overlap, the upper half of IDATA can only be accessed through indirect addressing mode, which hinders the effort to store local variables on stack. Thus some compilers will choose to avoid the upper half of IDATA, which reduces the available stack size to 223.

To improve the class 8051 architecture and avoid stack overflow, you can take the following measures:

- Expand the stack pointer. The stack pointer can be expanded to 16 bits. And the higher 8 bits of stack pointer can be stored in an unused SFR register.

- To avoid the awkwardness caused by SFR/IDATA overlap, the 16-bit stack pointer can work entirely on XDATA.

- However, the expansion of stack pointer has to work hand in hand with the C/C++ compilers. If the compiler does not fully support storing local variables with a 16-bit stack pointer, a less complicated scheme can be used, but only to store the functions return address on the stack and to statically store local variables for each function in XDATA. The price you pay for this static scheme is that your function cannot be re-entrant. In other words, recursive functions still have to use the stack to store their local variables for each invocation. As long as you can avoid recursive functions, the static allocation scheme can offer faster execution speed than the stack-based allocation scheme.

CISC Instruction Set

Like many processors born in the last century, 8051 carries a CISC instruction set. When you are trying to do a RISC implementation for a CISC instruction set like this, you will face the following challenges.

Variable Instruction Length

8051's instructions are not the same length. The shortest instruction has one byte, while the longest has three. For the old 12T 8051, such variable instruction length may not be a big deal, since the instruction could be fetched one byte at a time and still be able to achieve the 12 clock machine cycle. However, for this 1T 8051, the instruction fetch unit is supposed to get one complete instruction out of code memory in each clock cycle. To keep up, multiple memory banks can be used to read instructions out of code memory in parallel. Thus the word length of code memory can be 8, 16, 24 or 32 bits. But the variable instruction length makes it nearly impossible to store all instructions align to the word boundary, no matter what word length you choose. The misaligned read will complicate the design of instruction fetch unit and have a negative impact on timing closure.

Multitudes of Addressing Mode

Like typical CISC ISA, the 8051 instructions allow a multitude of addressing modes. In other words, there are multiple ways to index and access memory. For 8051, the available addressing modes for data memory are (Ref [20]):

- *Direct addressing:* The memory address is given in the instruction, such as ADD A, 123.

- *Indirect addressing:* The memory address is given in register R0 or R1, such as ADD A, @R0.

- *Immediate data:* A constant data becomes part of the instruction, such as ADD A, #123.

- *Bit addressing:* Some of the addresses in IDATA or SFR can be accessed bit-wise, as specified in Ref [29]. This feature comes in handy for software to toggle the IO port, like using "CPL 129" to flip the bit 1 of port 0. However, the handiness comes at the expense of a more sophisticated hardware. To simplify the design, the bit addressing can be treated like regular logic operations, and use the read-modify-write approach to keep other bits from the same byte unaffected.

For code memory access, the available addressing modes are as follows (Ref [20]):

- *16-bit direct addressing:* The full 16-bit address comprises the second and third byte of LCALL or LJMP instruction, which can be loaded into the PC directly.

- *11-bit direct addressing:* The 11-bit address mode is used by AJMP and ACALL instruction. It is composed of the 3 MSBs in the first byte and the full 8 bits in the second byte. Thus the jump or function call is confined to a page (2KB).

- *Relative jump:* In this mode, an 8-bit offset (signed number) from the second or third byte is added to the address of the following instruction, which limits the jump range between -128 bytes and 127 bytes.

- *Indirect addressing:* Code can also be moved into Accumulator through MOVC A, @ A+DPTR or MOVC A, @A+PC. Due to the restriction of the Harvard architecture, this is probably the only way to move things from code spaces into data spaces.

As you can see, there are multiple ways to access the same memory address, which could cause data hazards and make it hard to detect data dependency between instructions. This is part of the reason why RISC ISAs restrict memory access to load/store instructions only.

With the multitude of addressing modes, it looks like you need dedicate at least two pipeline stages to memory access, with one for memory read and the other for memory write. Otherwise it would be extremely difficult to keep the pipeline flow constantly and achieve 1T throughput. (In fact, as I will demonstrate shortly, you might have to add a third stage to accommodate the weak drive strength of block RAM.)

Optimizing for More Frequently Used Instructions

There are total of 111 distinctive instructions in 8051 ISA, including all variants of addressing mode mentioned earlier. They can be divided into the categories shown in Table 10-4.

Table 10-4. 8051 Instructions

Operations	Instructions	Number of Instructions
Arithmetic (excluding MUL/DIV)	ADD, ADDC, SUBB, INC, DEC	21
Multiplication/Division	MUL, DIV	2
Logic	ANL, ORL, XRL, CLR, CPL, RL, RLC, RR, RRC	24
Data Move	MOV, MOVX, XCH, PUSH, POP	25
Code Move	MOVC	2
Bit	CLR, SETB, CPL ANL, ORL MOV	12
No Operation	NOP	1
BCD (Binary Coded Decimal)	DA, SWAP, XCHD	3
Branch	JC, JNC, JB, JNB, JBC, ACALL, LCALL, RET, RETI, AJMP, LJMP, SJMP, JMP, JZ, JNZ, CJNE, DJNZ	21

Although you can try to make everyone in Table 10-4 be an 1T instruction, this may be both unnecessary and unrealistic for a number of reasons:

- Some instructions are used less often than others. For example, doing math directly in decimal format (BCD) might seem archaic today. To speed such instructions up will bloat the execution unit and hamper clock rate improvement.

- It takes more than one clock cycle to complete branch instructions. Most of the branch instructions in Table 10-4 will cause control hazards since they can make decisions based on the result of prior instructions, and such results are only available at the last stage of pipeline, which will inevitably stall the pipeline if branch prediction is wrong. In high-end processors, dual instruction fetch units and dual execution units can be in place to speed up the branch instructions, which may be overkill for low-end MCUs like 8051.

- There are generally two kinds of computation jobs in this world: control and process. As their names suggest, microcontrollers are more suitable for control intensive jobs, while process intensive jobs might be a good fit for DSP (Digital Signal Processor) or FPGA. In fact, 8-bit MCUs are not cut out to do heavy-duty mathematics as they cannot move data as fast as their 32-bit counterparts for obvious reasons. Thus we probably don't have to make multiplication and division as 1T instructions, as in the grand scheme of things those two operations are less critical for 8-bit MCUs.

- Reading things out of code space will inevitably interfere with instruction fetching. Thus as a compromise, MOVC will not be a 1T instruction either.

Given all these reasons, the next section tries to make a RISC implementation based on a CISC ISA for 1T 8051, making only 75% of the instructions in Table 10-4 to be 1T. By compromising on the other 25%, you can greatly reduce the design size and increase the clock speed.

RISC Implementation

Notwithstanding the fact that 8051 has a CISC instruction set, you could still translate those instructions into micro operations and apply a RISC implementation on top of it. The same approach has been taken by Intel to keep its x86 instruction set alive, while pushing the clock rate into the realm of GHz. (Of course, advanced process nodes also played a big part in this regard.)

Pipeline Design

As illustrated in Figure 10-2, the pipeline can be partitioned into five stages:

- IF: Instruction Fetch

- ID1: Instruction decode, part 1

- MEM: Read on-chip data, including IDATA, SFR and XDATA

- ID2: Instruction decode, part 2

- EX: Execution unit, data is written back to on-chip data memory at the end of this stage

There are two ID stages in this pipeline design, because:

- The nature of block RAM has to be taken into account when implementing a 1T 8051 core in FPGA. For a read, an unregistered block RAM will have its data available in the next clock cycle after address is given. However, the unregistered output cannot drive as much logic as that of flip-flops, and the weak drive strength will bog down the maximum clock rate. Thus data has to be latched into registers before they can drive other logics, which is why ID2 stage is sandwiched between MEM and EX in Figure 10-2 to account for this extra latency.

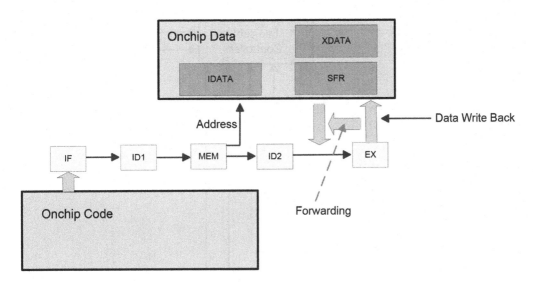

Figure 10-2. *Pipeline design for 1T 8051*

- To balance the workload, some less complicated instructions, such as RL A and RLC A, are decoded in ID2 instead of ID1. This logic retiming will let the whole design reach a higher clock rate.

In Figure 10-2, the result data are written back to memory at the end of the EX stage. And that is the only place where data is modified. Having a single point to modify memory will make interrupt handling a lot easier, because pipeline has to be flushed and reloaded at that moment.

To deal with data hazards (data dependency between stages), the output of EX stage can be forwarded to its input stage and bypass the output of IDATA/SFR/XDATA.

Pipeline stall will happen, especially with the indirect addressing mode, as shown in Listing 10-1.

Listing 10-1. Code Snippet for 8051 Indirect Addressing Mode

```
mov 104, #119

mov R0, #104

dec @R0
```

In Listing 10-1, the dec @R0 instruction will have to be stalled because R0 is not updated by mov R0, #104 until the EX stage is reached. So at the MEM stage, the read address for dec @R0 cannot be determined yet because at that time mov R0, #104 has only reached ID2. NOP will thus be inserted automatically until the new R0 value is available.

Peripherals and Interrupts

The classic 8051 has a very limited number of peripherals, such as two timers and one UART. However, the beauty of SOPC/FPGA is that you get to cherry pick the peripherals for your particular application and add them to your soft core 8051, as shown Figure 10-3. The control registers for those expanded peripherals can be added to SFR, utilizing those unused SFR addresses. For all peripherals, they can be connected to SFR through an ad hoc bus architecture, or with some extra effort, they can go through standard buses like Wishbone.

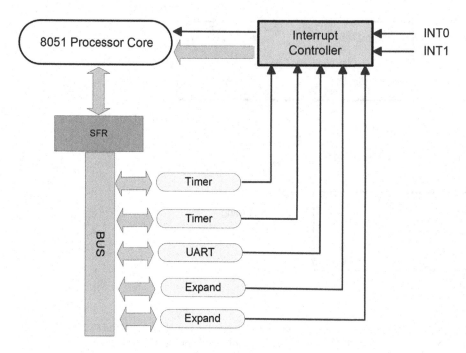

Figure 10-3. *Peripheral and interrupt for 8051*

An interrupt controller has to be in place to pass the peripherals interrupts and external interrupts (INT0/1) to the processor core, as illustrated in Figure 10-3. For 1T 8051, a vector interrupt controller can be used, which will pass vector addresses to the processor core directly.

Compilers and Debuggers

Have I ever mentioned that you need to prepare software for your processor? :-)

For 8051, in addition to the assembly code, there are also C/C++ compilers available, like those in Ref [31][32][33][34]. Among them, SDCC is probably the most popular one since it is the only open source C compiler available for 8051 so far. In addition to the C compiler, SDCC also offers SDCDB as a debugger. SDCDB's command-line interface is very close to that of GDB's, and it will run its code on a simulator called ucsim.

As I mentioned, you can use ROM monitor to debug your code. In fact, it is the approach that Keil C51 (Ref [32]) is taking, with its MON51 Target Monitor (Ref [35]).

However, for your 1T 8051, you probably could take a more advanced approach, like OCD (On Chip Debugger), as shown in Figure 10-4. In that case, the debugger will talk directly to the OCD through UART. And under the direction of the debugger, the OCD can collect information privy to the processor or set up hardware breakpoints if necessary.

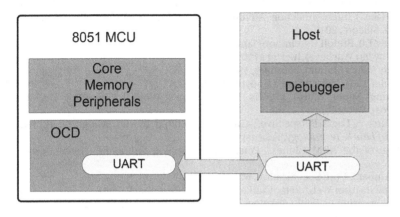

Figure 10-4. OCD (on-chip debugger)

Summary

Preliminary financial analysis is offered at the beginning of this chapter to show that the SOPC approach can make economic sense, as FPGA vendors keep churning out low-cost devices while the NRE cost for ASIC is skyrocketing.

The available choices for processor core and interconnections were thus discussed to demonstrate the diversity of SOPC. Among all the choices, 8051 stands out as a low-cost, popular solution since it is less burdened by patents and copyrights.

A RISC implement of 1T 8051 core was explored with depth, including the pipeline design, peripherals, interrupts, available compilers, and OCD (on chip debugger).

References

1. Schematic of Arduino (TM) UNO Rev 3, https://www.arduino.cc
2. 8-bit AVR Microcontroller with 8/16/32K Bytes of ISP Flash and USB Controller, ATmega8U2, ATmega16U2, ATmega32U2, ATMEL Corporation, 2010
3. FT232R USB UART IC Datasheet, Version 2.13, Future Technology Devices International Ltd, 2015
4. USB 3300 Hi-Speed USB Host, Device or OTG PHY with ULPI Low Pin Interface, Microchip Technology, Inc., January, 27, 2015
5. SN65LV1023A, SN65LV1224B Datasheet Rev E, 10-MHz To 66-MHz, 10:1 LVDS SERIALIZER / DESERIALIZER, Texas Instruments, December, 2009
6. Schematic of MP3 Shield, V15, Design by C. Taylor, N. Seidle, Revision by Byron Jacquot, Sparkfun, November 17, 2014
7. VS1053b Datasheet, VS1053b - Ogg Vorbis/MP3/AAC/WMA/FLAC/MIDI AUDIO CODEC CIRCUIT, Version 1.22, VLSI Solution (http://www.vlsi.fi), December 19, 2014
8. MAX9867 Ultra-Low Power Stereo Audio Codec, Rev 2, Maxim Integrated Products, Inc., June, 2010
9. Cortex-M1 Processor (http://www.arm.com/products/processors/cortex-m/cortex-m1.php)
10. List of 8051 devices available (http://www.keil.com/dd/chips/all/8051.htm)
11. ARM AMBA 5 AHB Protocol Specification, AHB5, AHB-Lite, ARM Limited Company, 2015
12. AMBA AXI and ACE Protocol Specification, AXI3, AXI4, and AXI4-Lite, ACE and ACE-Lite, ARM 2013
13. AMBA 4 ATB Protocol Specification, ATBv1.0 and ATBv1.1, ARM 2012
14. AMBA APB Protocol Version 2.0 Specification, ARM 2010
15. Avalon Interface Specifications, Altera Corporation, 12/10/2015
16. LEON3 Multiprocessing CPU core, LEON3 Product Sheet, Aeroflex Gaisler

17. WISHBONE System-on-Chip (SoC) Interconnection Architecture for Portable IP Cores, Revision B.4, OpenRISC, OPENCORES.ORG, Silicore, 2010
18. OpenCores Soc Bus Review, Rev 1.0, Rudolf Usselmann, January 9, 2001
19. The unofficial history of 8051, by Jan Waclawek (wek at efton.sk)
20. MCS-51 Microcontroller Family User's Manual, February, 1994
21. During of the Copyright (Circular 15A), United States Copyright Office, 2011
22. "Nintendo Says This Amazing Super Mario Site Is Illegal. Here's Why It Shouldn't Be.", by Timothy B. Lee, Washington Post, December 17, 2013
23. "Intel and the x86 Architecture: A Legal Perspective." Greg Tang, Edited by Ian Wildgoose Brown, JOLT Digest, *Harvard Journal of Law & Technology*, February 13, 2011
24. "Cambridge Calling: The Rise of the ARM Clones." Peter Clarke, Analog Editor, *EE Times Europe*, June 24, 2013
25. Amber Open Source Project, Amber 2 Core Specification, March 2015
26. "Student's ARM7 Clone Disappears from Web." Peter Clarke, *EE Times*, November 2, 2001
27. 8051 memory Spaces (8th revision), Paul Sokolovsky, GitHub contiki-os/contiki, May 2013 (https://github.com/contiki-os/contiki/wiki/8051-Memory-Spaces)
28. 80C51 Family Architecture, Philips Semiconductors, March, 1995
29. 80C51 Family Programmer's Guide and Instruction Set, Philips Semiconductors, September 18, 1997
30. 80C51 Family Hardware Description, Philips Semiconductor, December 1, 1997
31. SDCC Compiler User Guide, SDCC 3.4.0, March 22, 2014
32. C51 Development Tools, ARMKEIL Microcontroller Tools, http://www.keil.com/c51
33. IAR C/C++ Compiler Reference Guide, for the 8051 Microcontroller Architecture, Fifth Edition, April, 2011
34. 8051 C++ Compiler, The First Embedded C++ Compiler for 8051, MX, XA, and 251 Microcontrollers, Ceibo Inc. (http://ceibo.com/eng/products/cpp.shtml)
35. MON51 Target Monitor, ARMKEIL, (http://www.keil.com/c51/mon51.asp)

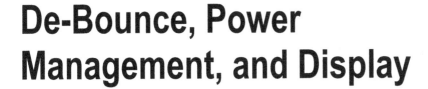

## CHAPTER 11

De-Bounce, Power Management, and Display

Your mobile phone has more computing power than all of NASA in 1969. NASA launched a man to the moon. We launch a bird into pigs.

—Tweet from @GeorgeBray (George Bray)

This chapter covers some miscellaneous hardware topics that are important to embedded systems, such as power-on resets, switch de-bouncing, power management, managing displays, and using touchscreens.

PoR (Power-On Reset) and Switch De-Bounce

No matter how smart your system becomes, one thing still remains a manual operation: Flipping the power switch. Two things have to be taken care of at the very moment when the system goes live:

- Make sure the system is reset properly so that it can have a fresh start.

- Make sure your switch does not "equivocate" when it is transitioning from one position to the other. In other words, it should not bounce. Or if it bounces, the bouncing should not cause unintended consequences.

Power-On Reset

People usually expect things to work right out of the box. That means the gadget they bought should work reliably from the moment when the power switch is flipped. From the standpoint of an embedded system engineer, it means:

- The system should be able to reliably generate a reset pulse during power on (a cold boot).

- After power on, the system should provide another mechanism for a warm boot. The usual arrangement is to have a reset pushbutton or a reset pinhole hidden on the back.

- As mentioned in previous chapters, a watchdog timer can also be used to initiate a warm reboot automatically.

Power-on reset can be done in the following ways:

- *Using an RC circuit.* A low-cost solution is to use RC circuit, as shown in Figure 11-1. The RC value can be adjusted to make sure the reset pulse is long enough.

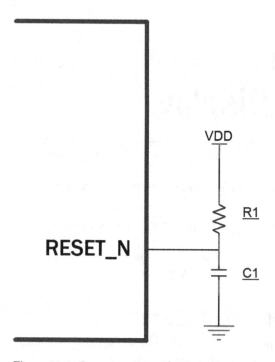

Figure 11-1. *Power-on reset with RC circuit*

- *Use a Schmitt Trigger.* The problem with an RC circuit is that it has a slow slew rate, which is prone to noise and can cause multiple state changes when reset_n is being de-asserted. An improvement can be made by introducing a Schmitt Trigger, as illustrated in Figure 11-2. The hysteresis nature of the Schmitt Trigger can turn the low slew rate pulse into sharp rise and fall. And the circuit is immune to noise as long as the peak-to-peak noise amplitude is less than the delta between Schmitt Trigger's V_{IH} and V_{IL} (Ref [2]).

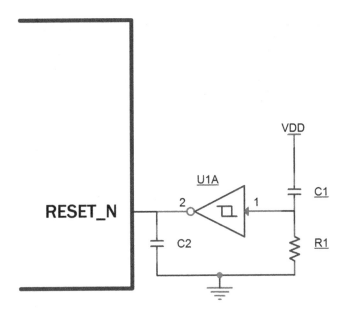

Figure 11-2. *Power-on reset with Schmitt Trigger*

- *Use power-on reset IC.* To reboot, we could introduce a pushbutton directly in Figures 11-1 or 11-2. The existing RC filter can be used as some sort of switch de-bounce circuit. However, a more reliable and compact solution is to use power-on reset ICs. Most of these kinds of ICs will support a power-on pushbutton with switch de-bounce. More advanced ICs will also include support for watchdog timers (Ref [3]).

Switch De-Bounce

Since switch de-bounce was mentioned multiple times previously, this section will serve to elaborate on it.

A switch (or a pushbutton) is a mechanical device. When a switch is flipped (or a pushbutton is pressed/released), its contact might bounce back and forth briefly before it settles at its new position. Such a contact bouncing is observed across board on almost all mechanical switches. The difference among them is just a matter of how long the bounce lasts. According to the first-hand statistics collected by Mr. Ganssle in Ref [4], the bounce time can be as little as sub-100 nanoseconds, or as long as hundreds of milliseconds. The bounce times for switch-open and switch-close can also be drastically different.

Obviously, such contact bouncing will electrically be translated into asynchronous glitches, and consequently introduce meta-stabilities into digital circuits. There are many ways to handle the bouncing. One is, of course, to use a dedicated circuit. Ref [4] has listed a few options for this. And if all else fails, you can always choose IC solutions (Ref [5]).

On the other hand, to save on BOM costs, the switch de-bouncing can also be implemented by firmware, FPGA, or CPLD. If it is done in firmware, the firmware will typically use a timer to read the corresponding IO pin for a consecutive number of intervals. If the reading is stable during the sample window, it can then be deemed a valid input. Ref [4] has a more detailed discussion about this firmware solution.

The downside of using firmware is that the timer interrupt it has introduced needs to be carefully integrated with the rest of the firmware. Since every embedded system is unique and carries its own idiosyncrasies, the firmware de-bouncer has to be done on a case-by-case basis.

Now, if there is already a CPLD or FPGA on board, adding the switch de-bouncer into them adds zero BOM cost. And a digital de-bouncer should be more stable and universal than its firmware counterpart.

As illustrated in Listing 11-1, after going through a shift register for clock domain crossing, the input will be used to control a counter. Any rising or falling edge on the input will reset the counter. The input will be deemed stable and valid if the counter can reach a predefined sample window size. In Listing 11-1, the sample window size is determined by the parameter TIMER_VALUE in a number of clock cycles. To make it more generic, TIMER_VALUE can be derived from a de-bounce interval (in milliseconds) and clock period when the module is being instantiated.

Listing 11-1. Implement Switch De-Bouncer in System Verilog

```
module switch_debouncer #(parameter TIMER_VALUE = 100000) (
        input wire clk,
        input wire reset_n,
        input wire data_in,
        output logic data_out
);

  logic unsigned [$clog2 (TIMER_VALUE) - 1 : 0]           counter;
  logic unsigned [3 : 0] data_in_sr;

  always_ff @(posedge clk or negedge reset_n) begin
      if (!reset_n) begin
              data_out      <= 0;
              counter       <= 0;
              data_in_sr    <= 0;
          end else begin
              if (data_in_sr [$high (data_in_sr)] != data_in_sr [$high
                              (data_in_sr) - 1]) begin
                  counter  <= 0;
                  data_out <= 0;
              end else if (counter == (TIMER_VALUE - 1)) begin
                      data_out <=   data_in_sr [$high (data_in_sr)];
              end else begin
                      counter  <=   counter + 1;
              end

              data_in_sr <= {data_in_sr [$high (data_in_sr) - 1 : 0], data_in};
          end
    end

endmodule : switch_debouncer
```

Power Management

The number of power domains in an embedded system often correlates to its complexity. Those tricks mentioned earlier might be good enough for small-scale systems when they are powered by external DC supply, with adequate ventilation. However, for large-scale systems or handheld applications, power management becomes a project by itself due to the following reasons:

- Modern day ICs, especially those System-on-Chip ICs, have multiple power rails for core, IO, etc. They have to be powered on with a pre-determined sequence. Violation of power-on sequence could cause latch-up[1] or even irreversible damage to the ICs.

- For battery powered devices, battery charge/recharge also needs to be closely monitored.

- To achieve maximum power savings, power domains will be selectively shut off depending on the circumstances. And for deep sleeping mode, the system could even switch to a much slower clock, such as 32KHz, to only have the house keeping job up while putting the rest of the circuit into sleep. Switching gears in and out of the sleep mode requires quite a lot of intelligence. This is especially true for TDMA-based handheld terminals. When going to sleep, the previous time tracking info will be saved and maintained by the 32KHz clock in coarse granularity. After wake-up and switch to the faster clock with fine granularity, the TDMA timing tracking should be restored as quickly as possible without going through the long reacquisition processing, which is always a challenging job for communication engineers.

One way to enlighten power management is to use microcontroller or CPLD for controlling the power-on sequence, as well as the power domain on/off. The plus side is that this approach is very flexible and expandable. The downside is, of course, the added BOM cost. In addition, since the voltage regulator and the microcontroller are all discrete components, they tend to take more spaces on the PCB.

Given these issues, IC vendors start to integrate the voltage regulators, the clock circuit, the battery charging circuit, and the microcontroller (or a simplified version of it) into one single IC called PMIC (Power Management IC). And more often than not, they will sell the PMIC as a companion chip to their System-on-Chip solution. To make things more compact, they would also integrate some other components, like Audio codecs or USB transceivers, into the PMIC. For many hand-held applications, the System-on-Chip solution seems to be the only way available for competitive products. So is the PMIC.

LCD Display

As the old saying goes, seeing is believing. The display industry has kept churning out new technologies one after another for the past 20 years. And those old bulky CRT displays have now been replaced by their flat panel successors. (Yes, that once almighty Sony Trinitron has fallen by the wayside.) Among those successors, LCD technology (with a wide variety of flavors) is the mainstay for embedded applications, although newcomers like OLED may become something big and solid in the future.

---

[1]In CMOS circuit, there are parasite BJTs formed by p well, n well, and the substrate. When these parasite BJTs are turned on by ESD or incorrect power-on sequence, a short circuit will be created. The creation of such short circuit is called latch-up. IC designers would usually put guard rings to prevent latch-up, but only to some extent.

Thanks to the abundance of new technologies, the display business is full of acronyms, such as LCD, LED, OLED, AMOLED, LTPS, IGZO, IPS, TFT, STN, ABN, and so on. The list goes on and on. A lot of these acronyms are deeply tied to the chemical processes they were produced with. Since I'm not a chemical engineer and the integration of display into embedded systems is what really matters, this book will only give a review of LCD display technology at 20,000-foot level, followed by a more detailed discussion on various display interfaces.

How LCD Works

The basic mechanism of LCD is to block or let through the light by changing the state of a liquid crystal. And unlike OLED, LCD by itself does not emit light.

LCD Cell

LCD stands for Liquid Crystal Display. In its natural form, the molecules inside the liquid crystal can have a twisted alignment instead of a uniformed one. This makes liquid crystal something between the solid state and the liquid state.

Because its molecule alignment is twisted, the liquid crystal can rotate the polarization of the light passing through it. In other words, the liquid crystal can act as an optical filter and add a polarization phase shift to the incoming light. On top of that, liquid crystal has another favorable nature. Under the influence of an external electrical field, liquid crystal's molecule alignment can be adjusted all the way to uniform alignment (untwisted). Light can then pass through it without any polarization phase shift. In other words, you can apply external electrical field to turn on or off the liquid crystal, like an optical switch (see Figures 11-3 and 11-4).

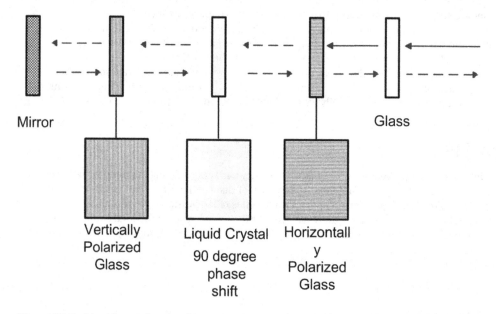

Mirror   Glass

Vertically   Liquid Crystal   Horizontall
Polarized   90 degree   y
Glass   phase   Polarized
                   shift   Glass

Figure 11-3. *Liquid crystal, twisted*

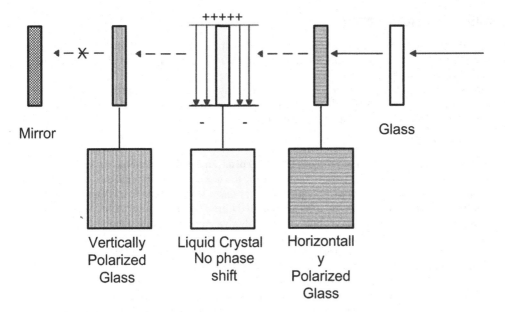

Figure 11-4. *Liquid crystal, untwisted*

With that in mind, a basic LCD cell can be formed as shown in Figure 11-3. It includes a regular glass, two polarized glasses (one is vertically polarized and the other is horizontally polarized), a mirror, and a liquid crystal sandwiched in the middle. The liquid crystal is carefully made so that it will have a 90-degree phase shift when it is in naturally twisted state. With the help of this 90-degree phase shift, light in Figure 11-3 is able to get through both polarized glasses that sandwich the liquid crystal and reflected back by the mirror. The eyes should see a bright cell in this case.

Now if you apply an electrical field to the liquid crystal and turn it into untwisted state, the light will stop at the second polarized glass it encounters, because the two polarized glasses are orthogonal to each other and the untwisted liquid crystal offers zero phase shift. The eyes should see a dark cell in this case.

In Figures 11-3 and 11-4, the light coming into the LCD cell is ambient light. There is no light source at the back of the cell. This kind of LCD display structure is called *reflective* display. They are often seen on battery powered calculators or digital stopwatches. (I was going to say "digital wrist watch" initially, but then I realized that it might get people confused with the Apple Watch.)

However, for poor-lighted settings, light has to come from the back of the LCD cell. Such an LCD is called *transmissive* display[2]. The backlight can be produced by florescent lamps or LEDs. And the latter is gaining in popularity with the market acceptance of LED TVs. However, you might not be able to see the image when the LCD is under direct sunlight.

As you can see, LCDs do not provide a light source, and thus backlighting is usually needed. And with the emergence of the OLED display, the necessity of backlighting can be eliminated completely since OLED can emit light by itself. The removal of backlighting is supposed to provide better picture quality than that of LCD. However, OLED has its own share of drawbacks, like low lifespan, low yield, high cost, etc. I believe engineers are still working hard to improve this technology. Inquisitive readers should keep a close watch on the latest industry news.

---

[2]As a trade-off between brightness, contrast, and dim environment, some LCDs called *transflective* display use both reflection and backlighting. Under sunlight, the backlight can be turned off, but it can be turned back on when the environment is dim. In this way, a transflective LCD can be viewed in both direct sunlight and lowlighting conditions.

Passive Matrix and Active Matrix

Displays are made of display elements. Those elements can be as simple as a few segments, like those seven segment digits on hand-held calculators. On the other hand, to display images or videos, the display elements usually have to form a pixel matrix.

Segment LCD is discussed later for its various programming interfaces. And for matrix LCDs, there are mainly two kinds of them: passive matrix and active matrix. They have different LCD cell structures, but both of them are addressed by multiplexing through the scan line.

- *Passive Matrix LCD (PMLCD)*

    As mentioned previously, to turn on or off an LCD pixel, an electrical field has to be applied or taken away. For passive matrix LCDs, the electrical field is realized through electrodes. As illustrated in Figure 11-5, assume a screen size of m rows and n columns. To display a frame, the rows are scanned through one at a time[3]. The n columns are selected simultaneously. By carefully choosing the code sequence for each column, a frame can be displayed scan line by scan line. However, unlike active matrix, passive matrix LCDs have no built-in mechanism to keep the electrical field. Once the pixel is not being actively selected, it will slowly restore to its off state. In other words, it takes some time (which could be on the order of 10^{-2} seconds according to Ref [8]) for the pixel change from one state to the other. If you do the scanning and configuration quickly enough, you can have an acceptable frame rate.

    Since the frame rate is usually between 60Hz and 200Hz, within the time when a pixel's state is being restored, the pixel can be addressed multiple times. Thus the effective voltage applied to the liquid crystal is the average of the signal voltage pulses over several frames (Ref [9]), and this RMS average could lead to very slow response time.

    The slow response time of passive matrix LCD makes it unsuitable for dynamic display content. In addition, with the increase of display density, cross talk will also increase due to the reduced electrode size and higher driving voltage (Ref [6]), which puts a constraint on the maximum number of scan lines that a passive matrix can have.

    The liquid crystal shown in Figures 11-3 and 11-4 has a 90-degree phase shift when it is in twisted state. Passive matrix made by these type of cells is called TN (Twisted Nematic). With the improvement of technology, TN passive matrix has been largely replaced by STN for better display quality[4]. The extra letter "S" in STN stands for "Super" and the STN can offer a phase shift from 180 degrees to 270 degrees. Today most passive-matrix LCDs on the market are STN type or its derivatives (CSTN, DSTN, FSTN, etc.). Like I said, those acronyms are deeply tied to the corresponding chemical process, and this book will not go any further on the chemical side.

---

[3]With the help of orthogonal functions, it is also possible to select multiple rows at the same time for passive matrix (Ref [8][10]).

[4]Note that TN and STN are merely referring to the liquid crystal cell structure while the passive or active matrixes mainly refer to the way how electrical fields are applied and kept. For STN, it is commonly used on passive matrix. But for TN, there are also TN-active-matrix LCDs being manufactured.

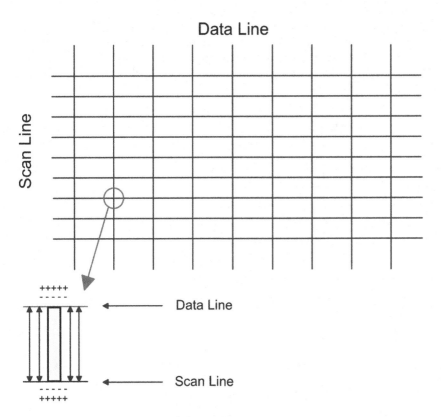

Figure 11-5. *Passive matrix addressing*

- *Active Matrix LCD (AMLCD)*

  A page can be borrowed from DRAM to improve the display quality of passive matrix. Instead of using electrodes to set up the electrical field, active matrix display uses a capacitor to keep the electrical field alive, as shown in Figure 11-6. Like the passive matrix, the columns of active matrix are also data lines that can be configured all at the same time. And each row is a scan line that is selected one by one. In active matrix, the scan line is used to control the switch that charges the capacitor in each LCD cell.

  Using a switch can greatly reduce the cross talk, and it also allows the voltage to be finely controlled. Adopting capacitors to keep the electrical field means a faster liquid crystal mixture can be used to reduce response time. With all those improvements, active matrix LCDs can have higher contrast, faster response time, wider viewing angle, more gray scales, and bigger screen size than their passive matrix counterparts. Of course, the down side is the added complexity and higher price relative to passive matrix.

  From the standpoint of manufacturing, there are multiple ways to make the switch. However, almost all AMLCDs made today use TFT (Thin Film Transistor) for switching. Thus manufacturers often use TFT LCDs to refer to active matrix.

The TFT LCD is constantly reinventing itself, and manufacturers keep churning out acronyms for new technologies, like IPS, IGZO, LTPS, etc. Those acronyms are mainly related to the chemical material used to make the TFT or the process associated with them, and each has its pros and cons. Again, this book will not go any further on the chemical side.

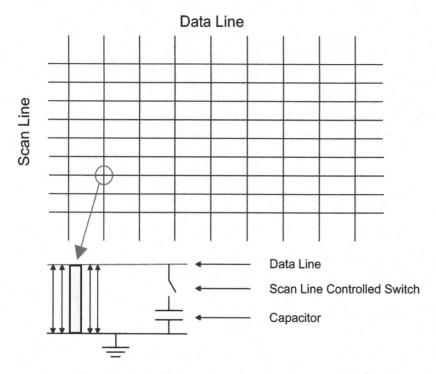

Figure 11-6. *Active matrix LCD*

LCD Programming Interfaces

Depending on the level of integration, LCD components can present different interfaces to engineers who program them. At the fundamental level, LCDs, whether they are segmented, passive matrix, or active matrix, all boil down to something sandwiched between two glass plates (substrates). A component like this is called LCD glass or LCD panel. LCD glasses only include the LCD cells and the glasses plates (with some metal or plastic casing maybe), plus all the necessary signals as pads around the package.

LCD glass by itself does not have much intelligence. If an LCD class or panel is chosen to be a component in the embedded system, the system designer should also throw in a driver IC to control the LCD glass. In such a setting, there are two ways to place the driver IC:

1. *Conventional surface mount*

   In this setting, the LCD glass is mounted to a PCB through some fine pitched connectors called ZEBRA connectors (or DIP connectors for some segmented LCDs). The driver IC is thus placed on the PCB and is hooked up to the LCD glass through the connector.

2. *COG (Chip on Glass)*

   In this case, for the two glass plates that sandwich the LCD cells, the bottom one is now extended out. The driver IC is placed on the extended part by some chemical seal. By doing this, COG can offer more flexibility in sizing as well as reduced cost. And with this flexibility, manufacturers can also make customized designs per request.

Instead of offering LCD glasses or LCD panels alone, LCD manufacturers often integrate the driver IC with the LCD glasses (conventional Surface Mount or COG), seal them into one package, and call it an LCD module[5]. LCD modules usually provide a microprocessor-friendly interface and can be connected directly to the main board through flex cable or ribbon cable.

Given all that, let's take a look at those interfaces at close range.

Segment LCDs

For segment LCDs, the system designer has two choices: Either she could buy an LCD module if there is one available, or she could buy an LCD glass and use a microcontroller as the driver IC. The latter will cut down the BOM cost, and LCD glass can also be custom made.

Assume an LCD glass is used and you need to find a way to control it. Although those segments in an LCD class can be in any shapes (dot, line, logo, patterns, etc.) and can be arranged in any layout form to suit the UI requirement, it can still be treated as a small scale passive matrix for control purposes. As demonstrated in Figure 11-7, the rows are called back plane (BP) or common and the columns are called front plane (FP) or segment line. Each segment is sandwiched between a BP and an FP.

---

[5]I'm not sure if there is an industry standard on those tech terms. But my impression is that LCD modules are often more advanced than LCD panels.

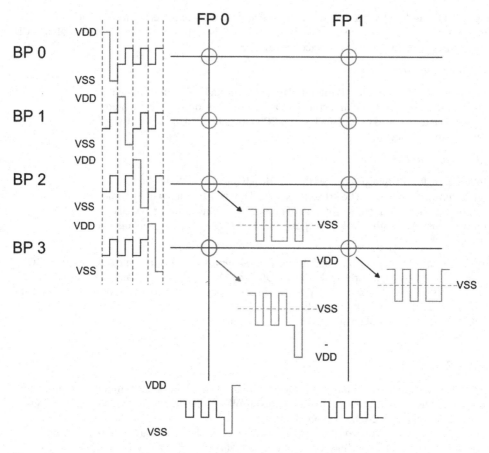

Figure 11-7. *Driving waveform, 1/4 duty cycle, 1/3 bias, eight segments*

As I previously mentioned, the rows are multiplexed by activating only one row at a time. One of the headaches with liquid crystal is that its drive voltage has to be DC free; otherwise the LCD's lifespan will be significantly reduced. To achieve that, a certain waveform has to be conceived. In other words, the inactive rows should also send out AC waveforms in certain formats, as demonstrated in Figure 11-7.

LCD glass manufacturers will usually tell you two things in their spec: duty cycle and BIAS. If the duty cycle is $1/N$, it means the LCD has N back planes. And if the BIAS is $1/S$, it means the maximum number of voltage levels that can be used is $(S+1)$, namely

$$\frac{VDD}{S} * i, (i = 0, 1, \ldots, S)$$

Here S is related to N. Since the rows are multiplexed, the state of the liquid crystals is determined by the RMS voltage applied on them. The screen contrast ratio is largely determined by the ratio between the RMS voltage of on-segments to that of the off-segments. By carefully choosing S to match N, the screen contrast ratio can be maximized. Ref [14] offers a table that lists the commonly used match between duty cycle and BIAS levels.

A sample waveform of 1/4 duty and 1/3 BIAS is shown in Figure 11-7. This sample waveform is called type A waveform, which keeps DC free in every single frame. But it has a lot of rising and falling edges that could lead to big side lobe in its spectrum. If side lobe is a concern, type B waveform (Ref [14]) can be used as an alternative. It will reduce the side lobe by keeping DC free for every two frames only.

Once the waveform is determined, the rest of the work is just to toggle the IO pins. First of all, the IOs have to be toggled fast enough to maintain a reasonable frame rate (> 30Hz) to avoid flickering. Secondly, the number of IOs is proportional to the total number of segments. Take an LCD with 4 commons and 120 segments for example—it needs 4 pins for commons (BPs) and 120/4=30 pins for segment lines (FPs).

Fortunately, there are lots of MCU vendors that include built-in LCD controllers in their products (Ref [15][17]). That makes the multiplexing control much easier. For BIAS that equals 1/2, general-purpose IO can be used with the help of external voltage divider, as shown in Ref [16]. And if the number of segments is small and only one common plane exists, a static control method can be used with each IO pin controlling one segment only. Under such circumstances, the BIAS is 1 and only two voltage levels are needed, so a general-purpose IO will suffice without any external voltage dividers.

VFD Display

Although this chapter is for LCD displays, VFD will be briefly mentioned for the sake of completeness, since VFD can be a good alternative to segment LCD.

VFD stands for Vacuum Fluorescent Display. It borrows a page from CRT's playbook. Like the CRT display, a VFD is made up of a filament as the cathode, a grid as the switch, and an anode coated with phosphor. For the on state, electrons emitted from the filament will pass through the grid, hit the anode phosphor, and light up. Conversely, the grid will stop the electrons from reaching the phosphor when the VFD is off (Ref [18]).

VFD trumps LCD in brightness, contrast, and viewing angle. And it is capable of operating at very low temperatures. Similar to LCD, VFD glasses can also be custom-made. Thus car manufacturers tend to use VFD on the dashboard (like the low oil sign or the engine light). The downside of VFD is its cost and power consumption. So it might not be the best option for hand-held applications, but it is still a popular choice for AC powered devices, such as microwave ovens or alarm clocks.

As for the programming interface, my takeaway is that VFD shifts a lot of work from software to hardware. For software, the concept of multiplexing versus static control still applies. However, VFD does not need a DC free control waveform, which makes the programming more straightforward. But at the hardware level, VFD routinely requires a high voltage (like 12 volt) supply to power the anode and grid, and it also demands an AC supply for the filament, thus extra driver circuit has to be in place to act as an inverter.

Character LCD Module (Dot Matrix LCD)

LCD vendors routinely offer LCD modules with predefined font ROMs. The font ROM normally includes the alphabet for western language or Japanese kana. Occasionally, the ROM will also include some predefined graphic pattern as an extension. Most LCD modules of this type are made of monochrome passive matrix LCD.

The standard size of character LCD ranges from 1 x 8 to 4 x 40. Its interface can be serial (like SPI) or parallel bus. If parallel bus is used, more often than not, it will be compatible with the Hitachi HD44780 LCD Controller, with an 8-bit data bus (Ref [19]). Since this interface has been universally adopted by the whole industry, application notes and open source drivers are plentiful (Ref [20][21]). And it is generally unnecessary to start from scratch without looking for an available driver first.

TFT Graphic LCD Module

Depending on the vendors, there are main two kinds of interfaces when it comes to programming a graphic LCD:

1. *MPU interface*

   This generally means a frame buffer has been integrated into the LCD module, and the LCD module will present itself as a SRAM to the controlling microprocessor. The interface can be a parallel bus of Intel 8080 or Motorola 6800 type, as mentioned in previous chapters.

2. *RGB interface*

   To reduce BOM cost and offer more flexibility, vendors will also offer an RGB interface with the absence of frame buffer. The RGB interface comprises the following signals:

   - VSYNC (Vertical Sync) and HSYNC (Horizontal Sync)

     The VSYNC is the sync pulse for row scan line, while the HSYNC is the sync pulse for each column. By default, the whole screen is scanned from the upper-left corner to the lower-right corner.

   - RGB data

     Aligned with each HSYNC pulse is the RGB data. The bit-width for each color component (R/G/B) varies from 4- to 8-bit.

   - Data enable

     Due to the real-time scan requirement, an external driver has to be present to generate the sync control and data signals. And in this case, an FPGA can be chosen to do the job, bridging between an LCD module of an RGB interface and a microprocessor.

Backlight Control

The brightness of an LCD screen can be adjusted by turning its backlight power on and off under a certain ratio. In other words, a PWM (Pulse Width Modulation) controller can be used to adjust the backlight.

There are two critical parameters that need to be determined by PWM: *switching frequency* and *duty cycle*. The switching frequency should be kept high (more than 200Hz or even in the range of KHz) to avoid possible flicking, and the duty cycle is usually proportional to the light intensity.

Implementation-wise, most microcontrollers have built-in PWM controllers. If not, a CPLD or FPGA can be used instead.

Using Touchscreens

Touchscreens often work hand in hand with LCD displays. The general idea of a touchscreen is to put a see-through layer on top of the LCD display[6], along with some sensors. When the screen is touched, the sensors help to locate the corresponding coordinates.

There is a plethora of sensor technologies that exist today for touchscreens, such as resistive sensor, capacitive sensor, SAW (Surface Acoustic Wave), infrared, etc. The industry has never stopped looking for new technologies. Since I'm not a sensor engineer and the integration into embedded systems is what really matters, I will take the same approach as I did for LCD: I give a review at a high level, followed by the discussion about the programming interface. And for the case of touchscreen, only the resistive and capacitive technologies will be discussed due to their popularity.

Resistive Touchscreens

Resistive touchscreen is often the preferred choice for low-cost applications. In its simplest form, a resistive touchscreen is made of two resistive sheets stacked together, with two wires on the bottom layer and the other two on the top, as illustrated in Figure 11-8. Gaps exist between those two layers and there is no direct contact between them under normal circumstances. However, when pressure is exerted upon by fingers or other hard objects, the top layer will bend over and touch the bottom layer, causing electrical conduction.

As shown in Figure 11-8, there are electrodes on the edge of both layers. At the bottom layer, the resistive strips go horizontally, while for the top layer, the resistive strips go vertically. To locate the coordinates of touch point, two steps have to be taken:

- Apply voltage to X+ and X- in Figure 11-8 and take the voltage from the Y axis. Since now the two layers are in contact, the whole thing acts like a potentiometer. From the voltage reading from Y axis, the X coordinate can be linearly determined.

- By the same token, apply voltage between Y+ and Y- and read the voltage from the X axis, which can help to determine the Y coordinate.

To measure the voltages in 4-wire touchscreens, two ADC channels are needed (one for X axis, the other for Y axis). Fortunately, there are microcontrollers out there that have ADC channels built-in, which can support the 4-wire touchscreen.

For large touchscreens, four additional wires (X+ sense, X- Sense, Y+ Sense, and Y- Sense) can be added as sensor points to help the calibration. And with enough ADC channels (four ADC channels for 8-wire touchscreen), both 4-wire and 8-wire touchscreens can be supported by using microcontrollers (Ref [22]).

---

[6]This is not always true, as sensors can also be integrated into the LCD display, such as is the case with the in-cell technology.

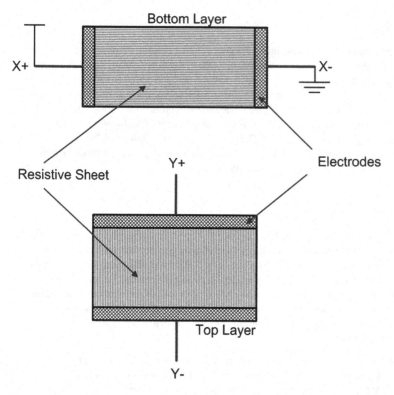

Figure 11-8. *Resistive touchscreen, four wires*

The biggest headache with 4-wire or 8-wire solutions is that the top layer is also used as a potentiometer. And in order for the touchscreen to work, the top layer has to be pressed hard to bend over and touch the bottom layer. Overtime, the hard-press will wear and tear the resistive coating, affecting the accuracy of coordinates' reading.

To improve the lifespan of this touchscreen, the 5-wire solution was conceived, as illustrated in Figure 11-9. At its bottom layer, electrodes are moved to the four corners. Voltage can still be applied vertically (from the top two electrodes to the bottom two) or horizontally (from the left two electrodes to the right two). The top layer is only used for measuring the voltage at the touch point.

In the 5-wire solution, the top layer is only used for probing the voltage at the touch point, so the wear and tear of it has less impact on the coordinates' reading. The touchscreen's lifespan can thus be enhanced. A microcontroller solution is also available for the 5-wire touchscreen (Ref [23]).

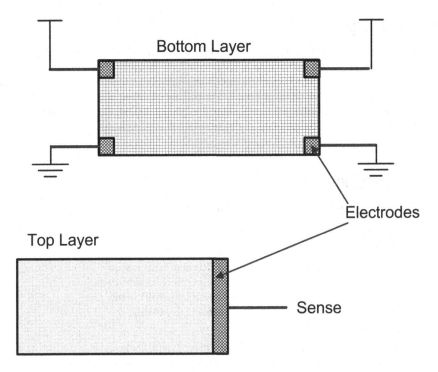

Figure 11-9. *Resistive touchscreen, five wires*

As mentioned earlier, the top layer has to be pressed hard to make the contact with the bottom layer. Thus, any hard object (finger, stylus, or even pencil with rubber cap) will work with the resistive touchscreen, but only one touch point can be detected and located. This makes it generally unsuitable for gesture recognition[7].

Capacitive Touchscreen

Capacitive touchscreen relies on the change of electrical fields to detect and locate the touch point. Compared to resistive touchscreen, capacitive use a glass surface. So no hard-press is needed, and the screen becomes more durable. However, in order to change the electrical field, conductive media, such as a human finger, must be introduced into the field, which means rubber headed pencil or wearing gloves may not work at all for capacitive touchscreen.

There are two kinds of capacitive touch technologies: surface capacitive technology and projected capacitive technology.

---

[7]This is not always true, though. With some extra effort, certain gestures can be recognized as well on resistive touch screens, as exemplified in Ref [24].

Surface Capacitive Technology

In surface capacitive technology, four electrodes are placed at each corner to create a uniformed electrical field. When the screen is touched by a finger, the electrical field is disturbed. By measuring the current flowing in and out of the electrodes, the touch point's coordinates can be precisely located.

Because it locates the touch point by comparing those currents in and out of corner electrodes, only a single point touch can be supported. With the omnipresence of smartphones these days, multi-touch and gesture recognition cannot be spared. To support multi-touch, projected capacitive technology has to be used.

Projected Capacitive Technology

There are two main sub-categories in the projected capacitive technology: self capacitance and mutual capacitance.

- *Self capacitance touchscreen*

  For self capacitance, only one electrode is used, and the RC value is measured between the electrode and ground. When a conductive object, such as a human finger, is introduced, the added capacitance from the human body will change the total capacitance value and affect the RC delay. By sensing the change, a touch point can be detected or located.

  The self capacitance touchscreen can be formed by one or two layers. If it is one layer, multiple electrodes dot the surface. If it is two layers, electrodes are formed as X or Y traces on each layer. Whether it's one layer or two layers, all electrodes are all measured individually.

  Since the electrodes are measured individually, ambiguity will take place for the two-layer structure. For this structure, when a two-finger touch happens, you can only determine the corresponding two rows and the two columns. However, that will give you four possible coordinates instead of two. But with some extra effort in software algorithms, gesture recognition, like the zooming gesture, might still be possible by detecting the moving direction of the touch points.

  To better support the multi-touch function, mutual capacitance can be used.

- *Mutual capacitance touchscreen*

  As its name suggests, two electrodes are used for mutual capacitance touchscreen. And the two-layer structure is adopted, with each electrode sitting on its own layer as X or Y traces.

  To detect and locate the touch, the screen is row scanned (or column scanned), which is very close to the way that LCD matrix works. The capacitance at each electrode intersection will be measured. Unlimited number of touch points can thus be supported by doing the scanning quickly enough. The scan results can be interpolated to further improve accuracy, at the expense of controller complexity.

On-Cell and In-Cell Touchscreen

For completeness, on-cell and in-cell touch are briefly mentioned in this section. Since those technologies are deeply tied to the manufacturing process, the discussion will be at a very high level.

The relentless drive for thinner smartphones is calling for new manufacturing processes. Traditionally, the smartphone screen is made of three layers from top to bottom:

- The glass cover at the top

- The capacitive touchscreen sandwiched in the middle

- The LCD panel at the bottom

To make the phone or tablet even thinner, some manufacturers choose to integrate the touch sensors into the glass layer, while others build the touch sensors directly into the LCD panel. To distinguish between them, the former approach is usually called on-cell touch, and the latter is called in-cell touch. Inquisitive readers can refer to the manufacturers' latest tech literatures for more information.

Programming Interface for Capacitive

As you can see from what was mentioned earlier, the capacitive touchscreen can be quite complicated, and a microcontroller alone may not be able to cut it. Most likely you are gonna choose a capacitive touch panel module, with built-in controllers to simplify the system design.

As illustrated in Figure 11-10, with the help of touch panel controller (TPC), the interface pin count between system MCUs and the touch panel module can be reduced to wake signal, interrupt, and some serial bus like I2C or SPI.

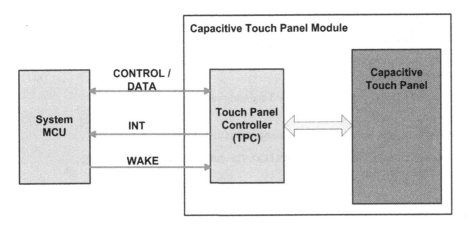

Figure 11-10. *Capacitive touch panel module*

The TPC can be an ASIC plus a microprocessor on the same PCB, or it can be an SoC with the processor core built-in. Now the detection capability is largely determined by the TPC alone, which is supposed to provide the following to system MCU:

- *Gesture detection*

  The TPC's capability is reflected by how many unique gestures it can recognize. Common gestures are move up/down, left right, zoom in/out, etc. A corresponding gesture code should be passed to the system MCU.

- *Locate touch point*

  The TPC's capability is also reflected by how many unique touch events it can handle. For each touch event, the position of the touch point should be located, along with the nature of the touch event, such as press down, release, firm contact, etc. An event code should also be passed on to the system MCU.

- *Interrupt to system MCU for new events*

- *Power saving*

With the assistance of TPC, the programming interface of the capacitive touch panel largely becomes interrupt handling and register read/write. Readers can refer to the module vendor's user manual for more information.

Summary

Miscellaneous hardware topics were discussed in this chapter, and they are summarized here:

- A reset circuit should be in place to guarantee a reliable power-on reset.

- The power-on sequence should be strictly followed.

- For reliability, the switch/button should be de-bounced, and the de-bouncing circuit can be implemented by FPGA or CPLD.

- An overview on LCD was given, followed by discussing various programming interfaces.

- LCD does not emit light by itself. Backlight control was also discussed.

- Touchscreens often work hand in hand with LCD. The corresponding programming interfaces were explored as well.

References

1. AN2790, TFT LCD interfacing with the high-density STM32F10xxx FSMC, Rev 2, STMicroelectronics, September, 2008
2. Using Schmitt Triggers for Low Slew-Rate Input, Application Note AC161, Actel Corporation, November, 2002
3. MAX6323/MAX6324 µP Supervisory Circuits with Windowed (Min/Max) Watchdog and Manual Reset, Maxim Integrated Products, Inc. 2010
4. "A Guide to Debouncing, or, How to Debounce a Contact in Two Easy Pages" (Rev 4). Jack Ganssle, March, 2014 (http://www.ganssle.com/debouncing.htm)
5. SC70/µDFN, Single/Dual Low-Voltage, Low-Power µP Reset Circuits, Maxim Integrated Products, 2012

6. Liquid Crystal Display (LCD) Passive Matrix and Active Matrix Addressing, Application Note (AN-002), Hitachi Europe Ltd, August, 2004

7. Fundamentals of Liquid Crystal Displays—How They Work and What They Do, Fujitsu Microelectronics American, Inc.

8. Liquid Crystal Display Drivers, Techniques and Circuits, Dr. David J.R. Cristaldi, Prof Salvatore Pennisi and Dr Francesco Pulvirenti, Springer International Publishing AG, 2009

9. Overview of the theory and construction of TFT display panels, SEQUOIA Technology, Ltd, December, 2003

10. Driving matrix liquid crystal displays, T N RUCKMONGATHAN, Raman Research Institute, *PRAMANA - Journal of Physics*, July 1999

11. Chip-on-Glass LCD Driver Technology, Well-proven Approach from NXP Reduces Medical System Design Costs, NXP July, 2012

12. AVR065: LCD Driver for the STK502, ATMEL Corporation

13. GDC130 Product Specification, DALIAN GOOD DISPLAY CO.LTD, March 11, 2008

14. AN52927, LCD Direct Drive Basics, Geethesh NS, Cypress Semiconductor, May 26, 2009

15. ATmega169A/PA/329A/PA/3290A/PA/649A/P/6490A/P 8-bit Atmel Microcontroller with 16/32/64KB In-System Programmable Flash DATASHEET, Atmel Corporation, July, 2014

16. AVR340: Direct Driving of LCD Using General Purpose IO, Atmel Corporation, September, 2007

17. MSP430xG461x MIXED SIGNAL MICROCONTROLLER, Texas Instruments, March, 2011

18. A guide to VFD operation, http://www.noritake-itron.com/SubPages/ApplicNotesE/vfdoperapn.htm

19. HD44780U (LCD-II) (Dot Matrix Liquid Crystal Display Controller/Driver, Hitachi Ltd, 1998

20. Interfacing a Hitachi HD44780 to a Silicon Laboratories C8051F120, Steve Dombrowski, Rensselaer Polytechnic Institute

21. Optrex 16207 LCD Controller Core, Altera Quartus II handbook, Vol. 5, Embedded Peripherals

22. 4-Wire and 8-Wire Resistive Touch-Screen Controller Using the MSP430, Neal Brenner, Shawn Sullivan, William Goh, Texas Instruments, November, 2010

23. AN10675, Interfacing 4-wire and 5-wire resistive touchscreens to the LPC247x, NXP Semiconductors, November, 2008

24. AD7879 Controller Enables Gesture Recognition on Resistive Touch Screens, Javier Calpe, Italo Medina, Alberto Carbajo, María José Martínez, Analog Dialogue Vol 45 June, 2011

25. *Designing Gestural Interfaces, Touchscreens and Interactive Devices.* Dan Saffer, O'Reilly Media, December, 2008

CHAPTER 12

Fixed Point Math

In rating ease of description as very important, we are essentially asserting a belief in quantitative knowledge—a belief that most of the key questions in our world sooner or later demand answers to 'by how much?' rather than merely to 'in which direction?'

—Exploratory Data Analysis, by John Wilder Tukey, 1977

I firmly believe that mathematics should be a big part of engineering. It is because every engineering project has risks associated with it, and knowing how to take calculated risks is what sets an engineer apart from the layman.

So this chapter is dedicated to math, or to be precise, fixed point math. Although technology advancements has made the FPU (Floating Point Unit) the norm to modern day CPUs, fixed point math still finds its place in those systems where low BOM costs and low clock rate are preferred. Compared to floating point math, fixed point math demands less complicated hardware at the expanse of more design effort from the engineers. This chapter addresses the commonly used fixed point functions.

Implementation-wise, there are two main ways to do fixed point math: CPU or dedicated logic. For the CPU approach, it can be a full-blown DSP (Digital Signal Processor), or it can be an application processor running at a high clock rate or with built-in support for SIMD instructions. Dedicated logic can be implemented through a FPGA, or it can be a hardware accelerator that becomes part of an SoC (System on Chip). Each approach has its pros and cons. And because there are two approaches for doing fixed point math, the code snippets presented in this chapter will also be in one of these two forms: C/C++ or Verilog/System Verilog.

Q Format

An essential part of fixed point math is knowing where the decimal point is. For a number with given word length, the true value it represents is determined by a scaling factor. Q format is the nomenclature that reflects both the word length and the scaling factor.

Q format is written as $Qm.n$, where m is the number of integer bits and n is the number of fractional bits. "m + n" is supposed to be equal to the word length. So for a 16-bit number, Q2.14 means it has 2 integer bits and 14 fractional bits.

However, note that the Q format is not always consistent across all technical literatures. In some literatures, "m + n" is equal is the word length minus 1. And sometimes the Q format is shortened to Qn to represent only the fractional part. But with a known word length, those ambiguities can be easily sorted out. And for this book, the notion $Qm.n$ will be used where "m + n" equals the word length.

© Changyi Gu 2016
C. Gu, *Building Embedded Systems*, DOI 10.1007/978-1-4842-1919-5_12

Two's Complement and Symmetric Range

For a signed binary number, the two's complement is the most commonly used representation format. An n bit two's complementary number can represent values from $(2^{n-1}-1)$ to (-2^{n-1}). And the addition and subtraction of a two's complementary number can be done directly by treating them as unsigned values.

One of the biggest problems with two's complement is its asymmetric range. As you can see, apart from zero, a two's complement has $(2^{n-1}-1)$ positive values and (2^{n-1}) negative values. In other words, the most negative value (-2^{n-1}) does not have a corresponding positive value. This would be a problem for functions like "absolute value" or "negation", because the most negative value (-2^{n-1}) will be mapped to itself in those operations if no extra care is taken.

Thus, my two cents is to eliminate the most negative value (-2^{n-1}) from very beginning. If input samples are taken from an ADC, those samples should be sifted through to bump the most negative value from (-2^{n-1}) to $(-2^{n-1}+1)$.

Basic Operations

The following basic operations are addressed by fixed point math:

- Rounding and truncation
- Saturation
- Negation
- Absolute value
- Addition, subtraction, multiplication, and division
- Sin and cos

Rounding and Truncation

For floating point math, the rounding is done by adding 0.5 to the LSB and having the result floored. That is, $Y = floor(X+0.5)$. Translated into fixed point math, it means adding 1 to the MSB of the fractional part and taking the integer bits. To deal with overflowing, one extra guard bit might be necessary, as illustrated in Figure 12-1.

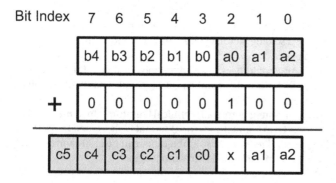

Figure 12-1. Fixed point rounding

In Figure 12-1, an 8-bit number is rounded to its lower three bits, which is carried out by adding 1 to bit 2 and taking the highest six bits (including the guard bit) in the sum. If five bits are desired after the rounding, a saturation operation has to follow.

In fact, in fixed point math, the rounding often comes after multiplication, after which it is often necessary to put the result back to its original word length. Listing 12-1 gives an example. An additional saturation operation can restore the result width back to DATA_WIDTH bits.

Listing 12-1. Rounding after Multiplication

```
wire  signed [DATA_WIDTH * 2 - 1 : 0]        x_times_y;
logic signed [DATA_WIDTH + 1 :0]             x_times_y_round;

assign x_times_y = data_in_x * data_in_y;

always_ff @(posedge clk) begin
     x_times_y_round <= x_times_y[DATA_WIDTH * 2 - 1 : DATA_WIDTH - 2] +
($size(x_times_y_round))'(1);
end

// And send x_times_y_round [DATA_WIDTH + 1 : 1] to saturation if necessary
```

Of course for truncation, it would be a lot more straightforward. An arithmetic right shift (shift right with sign bit extended) can be used in C/C++ code for doing the truncation.

Saturation

The rounding and truncation cut bits out at the LSB side, while the saturation operation does it on the MSB side.

For a saturation operation in its general form, assume the input data's bit-width is INPUT_DATA_WIDTH while the output data bit-width is OUTPUT_DATA_WIDTH. Overflow will be detected if the bits to be cut out are not the sign extension of the remaining bits. Its System Verilog implementation is shown in Listing 12-2.

Listing 12-2. Saturation Operation

```
wire signed [OUTPUT_DATA_WIDTH - 1 : 0] max_value_out;

assign max_value_out = {data_in[INPUT_DATA_WIDTH - 1], {(OUTPUT_DATA_WIDTH - 1)
{~data_in[INPUT_DATA_WIDTH - 1]}}};

always_ff @(posedge clk) begin
     if (enable_in) begin
          if (data_in[INPUT_DATA_WIDTH - 1 : OUTPUT_DATA_WIDTH] !=
            {(INPUT_DATA_WIDTH - OUTPUT_DATA_WIDTH)  {data_in[OUTPUT_DATA_WIDTH - 1]}})
               begin : over_flow_if
                    data_out <= max_value_out;
               end : over_flow_if
               else begin
                    data_out <= data_in [OUTPUT_DATA_WIDTH - 1 : 0];
               end
     end
end
```

More often than not, the saturation operation will be the step that follows an addition (or subtraction). In that case, the overflow can be determined by comparing the sign bit of the sum against that of the addends. And for addition, overflow will never happen if the two addends are in opposite signs. So the addition with saturation can be done in System Verilog, as shown in Listing 12-3.

Listing 12-3. Addition with Saturation

```
wire data_in_x_sign, data_in_y_sign;

wire signed [DATA_WIDTH - 1 : 0] x_plus_y;
wire signed [DATA_WIDTH - 1 : 0] max_value;

assign data_in_x_sign = data_in_x[DATA_WIDTH - 1];
assign data_in_y_sign = data_in_y[DATA_WIDTH - 1];

assign max_value = {data_in_x[DATA_WIDTH - 1], {(DATA_WIDTH - 1)
{~data_in_x[DATA_WIDTH - 1]}}};

assign x_plus_y = data_in_x + data_in_y;

always_ff @(posedge clk) begin
    if (enable_in) begin
            if ((data_in_x_sign != data_in_y_sign) ||
                (x_plus_y[DATA_WIDTH - 1] == data_in_x_sign)) begin
                    data_out <= x_plus_y;
            end else begin // over flow
                    data_out <= max_value;
            end
    end
end
```

And by the same token, the subtraction with saturation is shown in Listing 12-4.

Listing 12-4. Subtraction with Saturation

```
wire data_in_x_sign, data_in_y_sign;

wire signed [DATA_WIDTH - 1 : 0] x_minus_y;
wire signed [DATA_WIDTH - 1 : 0] max_value;

assign data_in_x_sign = data_in_x[DATA_WIDTH - 1];
assign data_in_y_sign = data_in_y[DATA_WIDTH - 1];

assign max_value = {data_in_x[DATA_WIDTH - 1], {(DATA_WIDTH - 1)
{~data_in_x[DATA_WIDTH - 1]}}};

assign x_minus_y = data_in_x - data_in_y;

always_ff @(posedge clk) begin
```

```
    if ((data_in_x_sign == data_in_y_sign) ||
        (x_minus_y[DATA_WIDTH - 1] ==  data_in_x_sign)) begin
            data_out <= x_minus_y;
    end else begin // over flow
            data_out <= max_value;
    end
end
```

Note that in all these saturation operations, the max value is composed by extending and flipping the sign bit. Although this makes the logic very compact, it will produce the max negative value as $\left(-2^{n-1}\right)$. In other words, it will produce an asymmetric data range between positive values and negative values, as explained earlier. Thus, you should be careful when those saturated values go through functions like "negation" or "absolute value".

And if an output with symmetric range is desired, the bit 0 of max value should be forced to be 1.

Negation

The negation of a two's complement can be done by inverting all of its bits, and add 1 to it afterward. However, the asymmetric range of a two's complement means the value $\left(-2^{n-1}\right)$ needs some special treatment, as shown in Listing 12-5 with System Verilog.

Listing 12-5. Negation

```
wire signed [DATA_WIDTH - 1 : 0]       data_neg;
wire                                   data_in_or, data_in_sign;

assign data_neg = (~data_in) + (DATA_WIDTH)'(1);

assign data_in_or = |(data_in[DATA_WIDTH - 2 : 0]);
assign data_in_sign = data_in[DATA_WIDTH - 1];

always_ff @(posedge clk, negedge reset_n) begin
    if (!reset_n) begin
            data_out <= 0;
            enable_out <= 0;
    end else begin
            data_out <= data_in_sign & (~data_in_or) ?
                        ({1'b0, {(DATA_WIDTH - 1){1'b1}}}) : data_neg;
            enable_out <= enable_in;
    end
end
```

In two's complement, the most negative value $\left(-2^{n-1}\right)$ will be represented as a 1 in the sign bit and all zeros for the other bits. In Listing 12-5, it will be mapped to $\left(2^{n-1}-1\right)$ instead of $\left(2^{n-1}\right)$.

Absolute Value

Although absolute value can be calculated by negating the negative values, a more compact solution is shown in Listing 12-6.

Listing 12-6. Absolute Value

```
wire signed [DATA_WIDTH - 1 : 0]      data_sign_ext, data_tmp;
wire                                  data_in_or, data_in_sign;
wire signed [DATA_WIDTH - 1 : 0]      abs_value;

assign data_sign_ext = {(DATA_WIDTH){data_in[DATA_WIDTH - 1]}};
assign data_tmp      = data_in ^ data_sign_ext;
assign abs_value     = data_tmp - data_sign_ext;

assign data_in_or = |(data_in[DATA_WIDTH - 2 : 0]);
assign data_in_sign = data_in[DATA_WIDTH - 1];

always_ff @(posedge clk, negedge reset_n) begin
    if (!reset_n) begin
        data_out <= 0;
        enable_out <= 0;
    end else begin
        data_out <= data_in_sign & (~data_in_or) ?
            ({1'b0, {(DATA_WIDTH - 1){1'b1}}}) : abs_value;
        enable_out <= enable_in;
    end
end
```

In Listing 12-6, sign bit is extended and used to exclusive-xor the original data. For positive values, such an operation has no effect, while for negative values this is identical to bit inversion. The sign bit is then subtracted from the exclusive-xor-ed result.

However, despite this neatness, the most negative value $\left(-2^{n-1}\right)$ still needs special treatment to map to $\left(2^{n-1}\right)$.

Unsigned Division

Up to this point, mathematicians haven't found a good way to do division on fixed point numbers. In other words, the number of cycles needed to obtain the quotient is always tied to the bit-width of numerator/denominator. To get the job done, three approaches will be discussed here.

- *Long hand division*

    This is the traditional way to do division by hand, as illustrated in Figure 12-2. Assume the numerator has a bit-width of DATA_WIDTH * 2 - 1, and the denominator has DATA_WIDTH bits. The long hand division is carried out by comparing the denominator against the numerator's higher DATA_WIDTH bits. A quotient bit of 1 is recorded if the denominator is smaller; otherwise, a bit 0 is recorded.

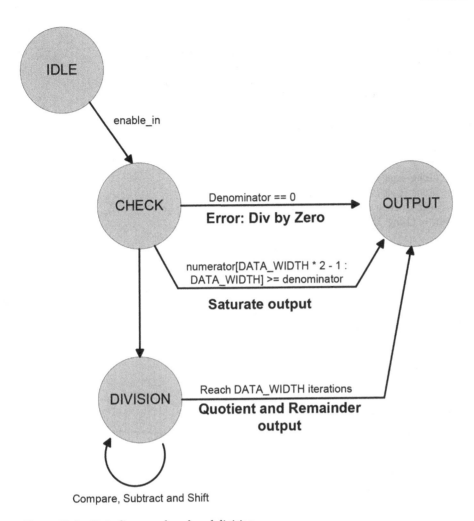

Figure 12-2. *State diagram: long hand division*

If a quotient bit 1 is recorded, the denominator (with zero extension) is then subtracted from the numerator. And the result is left shifted by 1 and used as the new numerator. If a quotient bit 0 is recorded, the numerator will be left shifted without any preceding subtraction. Repeat such a comparison (subtraction) and shift for DATA_WIDTH iterations to get the full quotient of DATA_WIDTH bits.

In addition to the division in DATA_WIDTH iterations, extreme cases like "divided by zero" and saturation also need to be taken care of to make the process bulletproof.

The long hand division is very straightforward, and it yields the most accurate result for the quotient and remainder. But it takes DATA_WIDTH iterations to get that. If speed is preferred over accuracy, the algorithm introduced in Ref [2] can be used.

- *Fast division*

    The so called Anderson-Earle-Goldschmidt-Powers algorithm was first applied
    on the ALU of IBM S/360 Model 91 mainframe computer. It is one of the best
    convergence-division algorithms that can yield satisfying results with much
    fewer number of iterations. Its C model is shown in Listing 12-7.

Listing 12-7. Anderson-Earle-Goldschmidt-Powers Algorithm

```c
int long_fast_div (int bitwidth, int n, long long Numerator, long long Denom, long long *Q)
{
    long long two = (long long)1 << bitwidth;
    long long one = (long long)1 << (bitwidth - 1);
    long long half = (long long)1 << (bitwidth - 2);
    long long quarter = (long long)1 << (bitwidth - 3);

    long long x, f, t;

    int i;

    if (Denom == 0) {
        return -1; // error
    }

//===== Normalize Denom
    while (Denom < half) {
            Denom = Denom * 2;
            Numerator = Numerator * 2;
    } // End of while loop

    while (Denom >= one) {
            Denom = Denom / 2;
            Numerator = Numerator / 2;
    } // End of while loop

//===== Main Loop
x = Numerator; t = Denom;

    for (i = 0; i < n; ++i) {

        f = two - t;

        x = x * f ;         x >>= (bitwidth - 2);
        ++x;        x >>= 1;
        t = t * f;
        t >>= (bitwidth - 2);
        ++t;
        t >>= 1;

        // printf ("i = %d, f = %llu, x = %llu, t = %llu\n", i, f,x,t);
    } // End of for loop
```

```
*Q = x >> bitwidth;

return 0;

} // End of long_fast_div()
```

To yield the best precision, both the numerator and the denominator should be normalized (shifted) so that the denominator will fall in the range of $[0.5, 1)$ in floating point. (The corresponding fixed point representation for .5 and 1 can be found in Listing 12-7.) If the algorithm is implemented by digital logic, the normalization part can be done with barrel shifters.

And in the main loop of Listing 12-7, x and t can be calculated in parallel. For a 16-bit number with four iterations ($n = 4$), the quotient produced by Listing 12-7 differs from that of long hand division's only by 0.002%. (i.e., most deviations come from the LSB only).

Due to the compactness of this algorithm, it can be easily implemented in FPGA. And it presents a good alternative to long hand division when throughput is prioritized over bit accuracy.

- *SRT division*

  In long hand division, the numerator is shifted by one bit in each step. To speed things up, people started to explore the possibility of handling multiple bits in one shot, and that's where SRT division came to the fore. The word SRT is from the first letter of its three inventors' names: Sweeney, Robertson, and Tocher (Ref [7][8][9][10]), who came up with the idea independently around the same time.

Now assume you would like to shift a numerator by two bits at a time (Radix 4 division). The mathematics works as the following:

In each step,

$$r_i = 4 \cdot r_{i-1} - q_i \cdot d, \qquad \qquad \text{(Equation 12-1)}$$

r_i is the remainder, q_i is the quotient digit, and d is the denominator. For the sake of convenience, we denote the partial remainder of $4 \cdot r_{i-1}$ by p.

Unlike the long hand division, the q_i in Equation 12-1 is allowed to take negative values:

$$-3 \le q_i \le 3$$

And you don't have to worry about the negative q_i at this point, because the strategy that SRT takes is to correct $q_0, q_1 \cdots q_{i-1}$ with a better q_i, and the final quotient is the accumulation of all of them, as demonstrated here:

$$Q = \sum_{i=0}^{n/2} q_i \cdot 4^{n/2-i}, \ (n = DATA\_WIDTH) \qquad \qquad \text{(Equation 12-2)}$$

Equation 12-2 will work as long as r_i converges. To guarantee convergence, it is imperative that the remainder for each step r_i stay between $-d$ and d. Consequently, you can deduce that:

$$-d \le r_i \le d \qquad \qquad \text{(Equation 12-3)}$$
$$-4 \cdot d \le p = 4 \cdot r_{i-1} = r_i + q_i \cdot d \le 4 \cdot d$$

From Equation 12-3, at each step, q_i can be chosen based on p and d. However, this could lead to two problems:

- For a given pair of p and d, there could be more than one value that satisfies Equation 12-3.

- Choosing a q_i value based on two parameters (p and d) complicates hardware design.

Preferably, you would like to come up with some criterion that:

- Be able to uniquely determine q_i.

- Be able to determine q_i with a simple algorithm. If q_i cannot be solely determined by p, at least it should only have an occasional reference to d.

With these requirements in mind, the Radix 4 SRT division can be carried out as the following:

1. Normalize d so that $\frac{1}{2} < |d| < 1$.

   For fixed point number with DATA_WIDTH=n, it means

   $$2^{n-1} \le |d| < 2^n .$$

2. Draw the plot of p versus d and partition it so that q_i can be determined with less effort.

   Figure 12-3 shows the p-d plot for 8-bit SRT division. (For the sake of simplicity, negative p values are omitted in Figure 12-3.) There are four straight lines that divide Figure 12-3 into separate regions. The equations for those four lines are:

   $$p = d$$
   $$p - 2 \cdot d$$
   $$p = 3 \cdot d$$
   $$p = 4 \cdot d$$

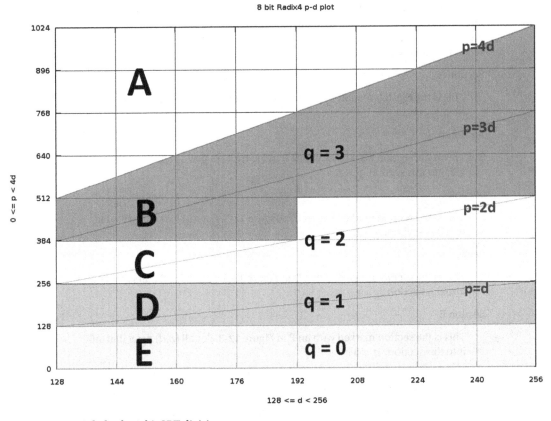

Figure 12-3. *p-d plot for 8-bit SRT division*

3. The corresponding q_i values that satisfy Equation 12-3 in each region are summarized in Table 12-1.

Table 12-1. *Valid q_i Values for Each Region in Figure 12-3*

Region	Valid q_i Value that Satisfies Equation 12-3
Below $p = d$	0, 1
Between $p = d$ and $p = 2 \cdot d$	1, 2
Between $p = 2 \cdot d$ and $p = 3 \cdot d$	2, 3
Between $p = 3 \cdot d$ and $p = 4 \cdot d$	3
Above $p = 4 \cdot d$	N/A, does not satisfy Equation 12-3

4. As p, d, and q_i all have to be integer values, the p–d plot inherently has an integer grid in its coordinates, as shown in Figure 12-3. By carefully analyzing Equation 12-3 and Table 12-1, the whole p–d plot can be divided into five sections as follows:

Section A

This is the top-left section in Figure 12-3. No meaningful values should exist in this section.

Section B

This is the section marked as with a B in Figure 12-3. For all (p,d) pairs that fall into this section, $q_i = 3$.

Section C

This is the section marked with a C in Figure 12-3. For all (p,d) pairs that fall into this section, $q_i = 2$.

Section D

This is the section marked with a D in Figure 12-3. For all (p,d) pairs that fall into this section, $q_i = 1$.

Section E

This is the section marked with an E in Figure 12-3. For all (p,d) pairs that fall into this section, $q_i = 0$.

As you can see, section D and E are rectangular. So only the MSB bits from p are needed to determine if (p,d) belongs to these sections.

Although section C is not a perfect rectangular, its shape is still very regular (unlike the grotesque shape of section B). Thus only the MSB bits from p and one bit from d is needed. (Remember that d has been normalized in Figure 12-3.)

So section B is the only one that has both valid values and an incongruous shape. However, since it is the only oddball among all the valid sections, its q_i can be determined when all else fails.

5. Thus, you now have criteria that are simple enough for hardware to implement. To prove it is correct mathematically, its corresponding Perl script is shown in Listing 12-8.

The Radix 4 SRT mentioned here allows q_i to be 3 or -3. Given the abundance of hardware resources in today's FPGA, the $3 \cdot d$ can be easily turned into $2 \cdot d + d$ without much impact on sizing or timing closure. So such an approach is good enough for FPGA and low-end microcontrollers.

For those processors that are targeting high-performance computations, q_i can be limited between -2 and 2, at the expense of a more complicated p–d plot (Ref [11]). The plus side is that it could speed things up a little bit. The flip side is the complicated p–d plot will turn into a big look-up table for more area. And mistakes in those tables could have dire consequences, both technically and financially, as Intel's Engineers can attest (Ref [12]). Google "pentium fdiv bug cost" to find out how much it cost Intel. :-(

Listing 12-8. SRT Radix 4 Division in Perl

```perl
sub SRT_radix4_div {;

    my($dividend, $divisor, $n)  = @_;
    my $i, $sign, $top4bits, $q, $normalize_shift;
    my $tmp, $tmp2, $final_Q, $quotient = 0;

    # normalize
    $normalize_shift = 0;
    for ($i = 0; $i < $n ; $i = $i + 1) {
        $tmp = 1 << ($n - $i - 1);
        $tmp2 = $divisor & $tmp;

        if ($tmp2) {
            last;
        } else {
            $normalize_shift = $normalize_shift + 1;
        }
    }

    $divisor = $divisor << $normalize_shift;

    if ($normalize_shift == $n) {
        exit(1); # divided by zero!
    }

    for ($i = 0; $i < ($n / 2 + 2); $i = $i + 1) {
        $sign = 1;
        if ($dividend < 0) {
            $sign = -1;
        }

        $top4bits = ($dividend) & (15 << ($n - 1));
        $top4bits = (($dividend) >> ($n - 1)) & 15

        if ($top4bits > 7) {
            $top4bits = $top4bits - 16;
        }

        if (($top4bits < 1) && ($top4bits >= -1)) {
            $q = 0;
        } elsif (($top4bits < 2) && ($top4bits >= -2)) {
            $q = 1;
        } elsif (($top4bits < 3) && ($top4bits >= -3)) {
            $q = 2;
        } elsif (($top4bits < 4) && ($top4bits >= -4)) {
            if ($divisor >= (6 * (1 << ($n - 3)))) {
                $q = 2;
            } else {
                $q = 3;
            }
```

```
        } else {
            $q = 3;
        }

        $q = $q * $sign;
        $dividend  = ($dividend  - $q * $divisor) * 4;
        $quotient = $quotient * 4 + $q;
    }
    $final_Q = $quotient    >> ($n + 2  - $normalize_shift
    return $final_Q;
}
```

Sin/Cos Table

There are many ways to calculate sin/cos functions. This section will discuss the approach of using look-up table. Later sections will address the CORDIC algorithm.

Due to the symmetric nature of sin/cos function, only 1/8th of a full circle (360 degrees) is needed for the table values. The rest of the circle can be deduced from those sin/cos values between 0 and 45 degrees, as shown in Listing 12-9.

Listing 12-9. Sin/Cos Table Using Look-Up Table

```
#define ANGLE_BITS 11
#define COS_SIN_BITS 16

const int cos_val [] = {
 32767,
 32767,
 32766,
  ...

 23312,
 23241,
 23170
};

const int sin_val[] = {
 0,
 101,
 202,

  ...

 23027,
 23098,
 23170
};
```

```c
void cos_sin_table (int index, int *cos_out, int *sin_out)
{
  int half_quad;
  int cos_base, sin_base;

  half_quad = index >> (ANGLE_BITS - 3);

  index &= (1 << (ANGLE_BITS - 3)) - 1;

  if ((half_quad % 2) == 0) {
      cos_base = cos_val [index];
      sin_base = sin_val [index];
  } else {
      cos_base = cos_val [(1 << (ANGLE_BITS - 3))- index - 1];
      sin_base = sin_val [(1 << (ANGLE_BITS - 3))- index - 1];
  }

  switch (half_quad) {
      case 0 :
              *cos_out = cos_base;
              *sin_out = sin_base;
              break;

      case 1 :
              *cos_out = sin_base;
              *sin_out = cos_base;
              break;

      case 2 :
              *cos_out = -sin_base;
              *sin_out = cos_base;
              break;

      case 3 :
              *cos_out = -cos_base;
              *sin_out = sin_base;
              break;

      case 4 :
              *cos_out = -cos_base;
              *sin_out = -sin_base;
              break;

      case 5 :
              *cos_out = -sin_base;
              *sin_out = -cos_base;
              break;
```

```
        case 6 :
               *cos_out = sin_base;
               *sin_out = -cos_base;
               break;

        case 7 :
               *cos_out = cos_base;
               *sin_out = -sin_base;
               break;

        default :
               *cos_out = cos_base;
               *sin_out = sin_base;
      }
}
```

In Listing 12-9, a sin/cos function with 2048 items is implemented, but only 256 sin/cos pairs are stored in the table. Out of the 11-bit index $\left(log_2^{2048} = 11 \right)$, the eight LSBs are used as a table index, while the top three MSBs are used as an index for the half quadrant (1/8th circle), which determines the sign or swapping of the table value. The total memory size is 16 * 2* 256 = 8K bits. And the latency is constant.

CORDIC Algorithm

In a mathematician's toolbox, the following elementary functions are often on his short list: sin, cos, arctan, amplitude, square root, exponentiation, logarithm, etc. Mathematicians tend to use Taylor serials to expand those elementary functions into polynomials and take it from there. Although fixed point math could follow the same approach, a good alternative that demands fewer resources is the CORDIC algorithm.

CORDIC stands for COordinate Rotation DIgital Computer. It was pioneered by Jack E. Volder in the 1950s to replace the analog navigation system on B-58 bombers (Ref [3]). At that time, the transistor was new to the industry and the clock rate was on the order of a few hundred KHz. So it was a very challenging job to calculate trigonometric functions and locate the coordinates for navigation in real time. But with the CORDIC algorithm, the problem was solved by using an iterative method with only addition/subtraction and shifting operations (plus a very small look-up table). Thanks to its ingenuity and compactness, the CORDIC algorithm was well received by the industry. And John S. Walther (Ref [5]) from HP Labs later generalized it and made it applicable to functions other than trigonometric ones.

Basic Ideas

In the CORDIC algorithm, there are basically three coordinate systems: circular, linear, and hyperbolic. All three coordinate systems follow the same curve equation:

$$x^2 + m \cdot y^2 = R^2$$

(Equation 12-4)

where m can take the value of 1 (for circular), 0 (for linear), or -1 (for hyperbolic).

Given an initial point (x_0, y_0) on the curve, the basic idea of CORDIC is to rotate the initial point iteratively by a serial of angles, designated as α_i $(i = 0,1,2,\ldots n-1)$. Depending on the goal of the rotation, there are two operation modes for each coordinate system: *vectoring mode* and *rotation mode*.

- *Vectoring mode*

    In vectoring mode, the goal is to force the y_n to be zero when $n \to \infty$. To achieve that, the sign of α_i has to change with y_i to make sure the iterations can converge.

- *Rotation mode*

    For rotation mode, a third variable z is introduced as $z_n = z_0 + \sum_{i=0}^{n-1} \alpha_i$,

    where z_0 can be any arbitrary value. And the goal of rotation mode is to force z_n to be zero when $n \to \infty$. To achieve that, the sign of α_i has to change with z_i to make sure the iterations can converge.

With three coordinate systems and two operation modes, there are total of six possibilities for generalized CORDIC algorithms. Their respective output functions are listed in Table 12-2.

Table 12-2. *CORDIC Output Functions*

Coordinate System	Vectoring Mode	Rotation Mode
Circular (m = 1)	$x_n = \sqrt{x_0^2 + y_0^2} / K_1$ $y_n = 0$ $z_n = z_0 + \tan^{-1}(y_0 / x_0)$	$x_n = (x_0 \cdot \cos z_0 - y_0 \cdot \sin z_0) / K_1$ $y_n = (y_0 \cdot \cos z_0 + x_0 \cdot \sin z_0) / K_1$ $z_n = 0$
Linear (m = 0)	$x_n = x_0$ $y_n = 0$ $z_n = z_0 + y_0 / x_0$	$x_n = x_0$ $y_n = y_0 + x_0 \cdot z_0$ $z_n = 0$
Hyperbolic (m = -1)	$x_n = \sqrt{x_0^2 - y_0^2} / K_{-1}$ $y_n = 0$ $z_n = z_0 + \tanh^{-1}(y_0 / x_0)$	$x_n = (x_0 \cdot \cosh z_0 + y_0 \cdot \sinh z_0) / K_{-1}$ $y_n = (y_0 \cdot \cosh z_0 + x_0 \cdot \sinh z_0) / K_{-1}$ $z_n = 0$

$K_1 = 0.607253$

$K_{-1} = 1.207497$

The rest of this chapter attempts to explain Table 12-2 in more detail.

Circular Coordinate System

Assume the angle of point (x, y) is θ. The circular coordinate system can be expressed by trigonometric functions as $x = R \cdot \cos\theta, y = R \cdot \sin\theta$. Now assume you rotate it by another angle of α, the new point will be

$$x' = R \cdot \cos(\theta + \alpha) = R \cdot \cos\theta \cdot \cos\alpha - R \cdot \sin\theta \cdot \sin\alpha = x \cdot \cos\alpha - y \cdot \sin\alpha$$

$$y' = R \cdot \sin(\theta + \alpha) = R \cdot \sin\theta \cdot \cos\alpha + R \cdot \cos\theta \cdot \sin\alpha = y \cdot \cos\alpha + x \cdot \sin\alpha$$

Thus

$$\begin{pmatrix} x' \\ y' \end{pmatrix} = \begin{pmatrix} \cos\alpha & -\sin\alpha \\ \cos\alpha & \sin\alpha \end{pmatrix} \cdot \begin{pmatrix} x \\ y \end{pmatrix} = \cos\alpha \cdot \begin{pmatrix} 1 & -\tan\alpha \\ 1 & \tan\alpha \end{pmatrix} \cdot \begin{pmatrix} x \\ y \end{pmatrix}$$

$$= \frac{\sqrt{(\cos\alpha)^2}}{\sqrt{(\cos\alpha)^2 + (\sin\alpha)^2}} \cdot \begin{pmatrix} 1 & -\tan\alpha \\ 1 & \tan\alpha \end{pmatrix} \cdot \begin{pmatrix} x \\ y \end{pmatrix}$$

$$= \frac{1}{\sqrt{1 + (\tan\alpha)^2}} \cdot \begin{pmatrix} 1 & -\tan\alpha \\ 1 & \tan\alpha \end{pmatrix} \cdot \begin{pmatrix} x \\ y \end{pmatrix}$$

For each iteration, if you intentionally set $\tan\alpha$ to be power of 2 (such as $\pm 2^{-i}$ for ith iteration), the multiplications will be turned into a shift operation. The part $\dfrac{1}{\sqrt{1 + (\tan\alpha)^2}}$ can then be pre-calculated as constants and stored in a look-up table.

Circular Vectoring Mode

As indicated in Table 12-2, the vector mode in the circular coordinate system will give you amplitude and *arctan*. Its fixed point C model is shown in Listing 12-10. (The constant CORDIC_SCALING_FACTOR is defined as a scaling factor for all fixed point numbers. And theta[i] can be pre-calculated as well.)

Listing 12-10. Vectoring Mode in Circular Coordinate System

```
#define CORDIC_MAX_ITERATIONS 30
#define CORDIC_SCALING_FACTOR (1 << 30)

int CORDIC_circular_vectoring_mode(int n, long long *pxo, long long *pyo)
{
  unsigned int theta[CORDIC_MAX_ITERATIONS];
  int i;
  int sigma;

  long long x,y;
  int z;
  long long x_tmp, y_tmp;
  double k = 1;

  x = *pxo; y = *pyo;
  z = 0;
```

```
for (i = 0; i < n; ++i) {

    theta[i] = floor((atan2 (1, 1 << i) * CORDIC_SCALING_FACTOR + 0.5));
    k = k / sqrt (1 + 1.0 / (((long long)1) << (2 * i)));

    if  (y >= 0)  {
         sigma = -1;
    } else {
         sigma = 1;
    }

    if (i > 0) {
         x_tmp = x - sigma * (((y >> (i - 1)) + 1) >> 1);
         y_tmp = (((x >> (i - 1)) + 1) >> 1) * sigma + y;
    } else {
         x_tmp = x - sigma * y;
         y_tmp = x * sigma + y;
    }

    x = x_tmp;
    y = y_tmp;

    z = z - theta[i] * sigma;

    printf ("i = %d, k = %lf\n", i, k);

} // End of for loop

*px0 = x;

return z;

}
```

For demonstration purposes, the pre-calculated number

$$K(n) = \prod_{i=0}^{n-1} \frac{1}{\sqrt{1+2^{-2i}}} \quad (n=1,2,3\ldots)$$

is crunched on each iteration in Listing 12-10, and the results are shown in Table 12-3.

Table 12-3. *Pre-Calculated Constant for Circular Coordinate Systems*

Number of Iterations n	K (n)
1	0.707107
2	0.632456
3	0.613572
4	0.608834
5	0.607648
6	0.607352
7	0.607278
8	0.607259
9	0.607254
10	0.607253
11	0.607253
12	0.607253
13	0.607253
...	0.607253

As you can see, the pre-calculated constant starts to converge on 0.607253 when the n becomes big, and that is the value K_1 (here the subscript '1 stands for m=1) referenced in Table 12-2.

As shown in Listing 12-11, the amplitude function can be calculated based on Listing 12-10.

Listing 12-11. CORDIC Amplitude Function

```
long long CORDIC_amplitude (long long x, long long y)
{
  long long amp;
  long long K1 = 0.607253 * CORDIC_SCALING_FACTOR;

  CORDIC_circular_vectoring_mode(CORDIC_MAX_ITERATIONS, &x, &y);

  amp = (K1 * x + (CORDIC_SCALING_FACTOR / 2))/ CORDIC_SCALING_FACTOR;

  return amp;
}
```

The *arctan* function can be calculated as shown in Listing 12-12.

Listing 12-12. CORDIC Arctan Function

```
int CORDIC_arctan (long long x, long long y)
{
  if ((y == 0) && (x == 0)) {
      return -1; // error condition
```

```
    } else {
        // scale for best precision
        while ( (abs(x) < (CORDIC_SCALING_FACTOR / 2)) &&
                (abs(y) < (CORDIC_SCALING_FACTOR / 2))) {
                x *= 2;
                y *= 2;
        } // End of while loop

        return CORDIC_circular_vectoring_mode (CORDIC_MAX_ITERATIONS, &x, &y);
    }
}
```

Note that the x and y are scaled up in Listing 12-12 to achieve the best precision.

Circular Rotation Mode

The rotation mode is similar to the vectoring mode, except its goal is to force z to zero. As indicated in Table 12-2, the rotation mode in circular coordinate systems should give you the combination of $\cos z_0$ and $\sin z_0$. And sin/cos values can be singled out and extracted by carefully choosing x_0 and y_0.

Listing 12-13 shows the fixed point C model for rotation mode in circular coordinate systems. The corresponding sin/cos functions (with examples) are shown in Listing 12-14. Compared to the look-up table method in the previous section, the CORDIC approach can achieve good precision with fewer resources and longer latency. And in order to let CORDIC algorithm converge, the input angle z_0 has to be kept between $-\pi/2$ and $\pi/2$.

Listing 12-13. Rotation Mode in Circular Coordinate System

```
long long CORDIC_circular_rotation_mode(int n, long long *px0, long long *py0,
                                        signed long long z0)
{
    unsigned int theta[CORDIC_MAX_ITERATIONS];
    int i;
    int sigma;

    long long x,y;
    long long z;
    long long x_tmp, y_tmp;

    x = *px0; y = *py0;
    z = z0;
    for (i = 0; i < n; ++i) {
        theta[i] = floor((atan2 (1, 1 << i) * CORDIC_SCALING_FACTOR + 0.5));

        if (z < 0) {
            sigma = -1;
        } else {
            sigma = 1;
        }
```

```
        if (i > 0) {
                x_tmp = x - sigma * (((y >> (i - 1)) + 1) >> 1);
                y_tmp = (((x >> (i - 1)) + 1) >> 1) * sigma + y;
        } else {
                x_tmp = x - sigma * y;
                y_tmp = x * sigma + y;
        }

        x = x_tmp;
        y = y_tmp;

        z = z - (long long)theta[i] * sigma;

        } // End of for loop

  *px0 = x; *py0 = y;
  return z;
}
```

Listing 12-14. CORDIC Sin/Cos Functions

```
long long CORDIC_sin (long long z)
{
  long long K1 = 0.607253 * CORDIC_SCALING_FACTOR;
  long long x = CORDIC_SCALING_FACTOR;
  long long y = 0;

  CORDIC_circular_rotation_mode(CORDIC_MAX_ITERATIONS, &x, &y, z);

  return ((K1 * y + (CORDIC_SCALING_FACTOR / 2))/ CORDIC_SCALING_FACTOR);
}

long long CORDIC_cos (long long z)
{
  long long K1 = 0.607253 * CORDIC_SCALING_FACTOR;
  long long x = CORDIC_SCALING_FACTOR;
  long long y = 0;
  CORDIC_circular_rotation_mode(CORDIC_MAX_ITERATIONS, &x, &y, z);
  return ((K1 * x + (CORDIC_SCALING_FACTOR / 2))/ CORDIC_SCALING_FACTOR);
}

//==============================================================================
// Example:
//
// z0 = floor((atan2 (1, 2) * CORDIC_SCALING_FACTOR + 0.5));
// cosz = CORDIC_cos (z0);
// sinz = CORDIC_sin (z0);
```

```
//
// printf ("cos = %f, sin = %f\n",
// (double)cosz / CORDIC_SCALING_FACTOR, (double)sinz / CORDIC_SCALING_FACTOR);
//
// cos = 0.894427, sin = 0.447214
//
```

Linear Coordinate System

For a linear coordinate system, its curve is a straight line with $x = R$ according to Equation 12-4. However, I found it much easier to understand by viewing x as the radius of a circle and y as the length of an arc, with the included angle as θ in radians. Figure 12-4 illustrates this idea for $R = 1$.

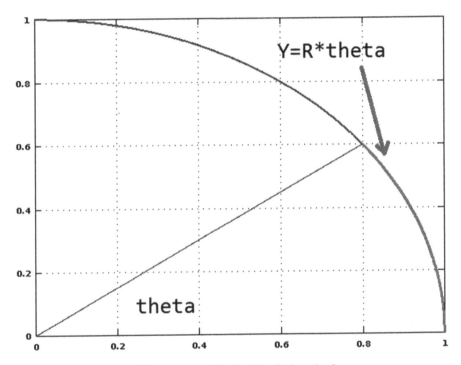

Figure 12-4. *Linear coordinate system, viewing Y as the length of arc*

Thus, a point (x, y) in a linear coordinate system can be expressed as $x = R$ and $y = R \cdot \theta$. Assume you rotate it by another angle of α, the new point will be

$$x' = x = R;\ y' = R \cdot (\theta + \alpha) = y + x \cdot \alpha$$

Namely,

$$\begin{pmatrix} x' \\ y' \end{pmatrix} = \begin{pmatrix} 1 & 0 \\ \alpha & 1 \end{pmatrix} \cdot \begin{pmatrix} x \\ y \end{pmatrix}$$

293

For each iteration, if you intentionally set α to be a power of 2 (in this case, it will be $\pm 2^{-i}$, $i=1,2,3,\dots$), the multiplication will become a shift operation. But unlike the circular coordinate system, there is no need to pre-calculate anything for the linear coordinate system.

Linear Vectoring Mode

The fixed point C model for vectoring mode in a linear coordinate system is shown in Listing 12-15. According to Table 12-2, this mode can do division, as illustrated in Listing 12-16.

Listing 12-15. Vectoring Mode in Linear Coordinate System

```
int CORDIC_linear_vectoring_mode (int n, long long x0, long long y0, int z0)
{
  int i;
  int sigma;

  long long x,y;
  int z;
  long long x_tmp, y_tmp;

  x = x0; y = y0;
  z = z0;

  for (i = 1; i <= n; ++i) {

      if (y >= 0)  {
          sigma = -1;
      } else {
          sigma = 1;
      }

      x_tmp = x;
      y_tmp = (((x >> (i - 1)) + 1) >> 1) * sigma + y;

      z = z - (CORDIC_SCALING_FACTOR >> i) *   sigma;

      x = x_tmp;
      y = y_tmp;

  } // End of for loop

  return z;
}
```

Listing 12-16. CORDIC Division Function

```
int CORDIC_division (long long x, long long y)
{
  if ((y == 0) && (x == 0)) {
      return -1; // error condition
  } else {
      // scale for best precision
      while ((abs(x) < (CORDIC_SCALING_FACTOR / 2)) && (abs(y) <
  (CORDIC_SCALING_FACTOR / 2))) {
                  x *= 2;
                  y *= 2;
      } // End of while loop

      return CORDIC_linear_vectoring_mode (CORDIC_MAX_ITERATIONS, x, y, 0);
  }
}
```

Note that just like Listing 12-12 for the *arctan* function, the *x* and *y* in Listing 12-16 are also scaled up to achieve the best precision.

The previous sections discussed long hand division, fast division, and SRT division. It would be interesting to compare them to CORDIC division. Here is an example to calculate 2490/3276 with a bit-width of 30:

$$2490 \div 3276 = 0.760073260073260$$
$$2490 / 3276 \times 2^{30} = 816122448.644689 \approx 816122449$$

(Equation 12-5)

So the long hand division will give you the quotient of 816122448, and the fast division in Listing 12-7 will also give the same answer with four iterations and a bit-width of 31. The CORDIC division in Listing 12-16 will present the result as 816122449, which is asymptotically equal to the value in Equation 12-5.

Compared to fast division or SRT division, CORDIC division holds no advantage on latency or precision. The only thing CORDIC division can do better is the resource usage since it requires no multiplier. However, given the fact that built-in DSP block is ubiquitous in modern day FPGA, fast division or SRT division becomes a more favorable choice over CORDIC division when it comes to FPGA implementation.

Linear Rotation Mode

The fixed point C model for vectoring mode in a linear coordinate system is shown in Listing 12-17. According to Table 12-2, this mode can do multiplication, as illustrated in Listing 12-18.

Listing 12-17. Rotation Mode in Linear Coordinate System

```
int CORDIC_linear_rotation_mode(int n,
                            long long x0, long long y0, long long z0)
{
  int i;
  int sigma;

  long long x,y;
  long long z;
  long long x_tmp, y_tmp;
```

```
      x = x0; y = y0;
      z = z0;
      for (i = 1; i <=n; ++i) {
          if  (z >= 0)  {
                  sigma = 1;
          } else {
                  sigma = -1;
          }

          x_tmp = x ;
          y_tmp = (((x >> (i - 1)) + 1) >> 1) * sigma + y;

          z = z - (long long)(CORDIC_SCALING_FACTOR >> i) *    sigma;

          x = x_tmp;
          y = y_tmp;

      } // End of for loop

      return y;
}
```

Listing 12-18. CORDIC Multiplication Function

```
#define CORDIC_MAX_ITERATIONS       30

#define CORDIC_SCALING_FACTOR      (1 << CORDIC_MAX_ITERATIONS)
#define CORDIC_SCALING_FACTOR_SQRT (1 << (CORDIC_MAX_ITERATIONS / 2))

long long CORDIC_mult (int x, int y)
{
  long long result;
  result = CORDIC_linear_rotation_mode (CORDIC_MAX_ITERATIONS + 1,
          (long long)x * CORDIC_SCALING_FACTOR_SQRT * 2, 0,
          (long long)y * CORDIC_SCALING_FACTOR_SQRT);

  result = (result + 1) >> 1;

  return result;
}
```

To make sure z would converge to zero, its original value $z0$ should be kept less than CORDIC_SCALING_FACTOR, thus the CORDIC multiplication in Listing 12-18 did some scaling and limited the x and y to be no more than $\sqrt{CORDIC\_SCALING\_FACTOR}$. For the particular case in Listing 12-18, the maximum x or y is 32767.

And for the case of Listing 12-18, it takes 31 iterations for CORDIC to do the multiplication, which is extremely slow. Given the fact that built-in DSP blocks are ubiquitous in modern day FPGA, CORDIC multiplication is rarely used in practical FPGA design.

Hyperbolic Coordinate System

Assume the hyperbolic angle of point (x, y) is θ. The hyperbolic coordinate system can be expressed as $x = R \cdot \cosh \theta$, $y = R \cdot \sinh \theta$. Now assume you rotate it by another angle of α. The new point will be

$$x' = R \cdot \cosh(\theta + \alpha) = R \cdot \cosh \theta \cdot \cosh \alpha + R \cdot \sinh \theta \cdot \sinh \alpha = x \cdot \cosh \alpha + y \cdot \sinh \alpha$$

$$y' = R \cdot \sinh(\theta + \alpha) = R \cdot \sinh \theta \cdot \cosh \alpha + R \cdot \cosh \theta \cdot \sinh \alpha = y \cdot \cosh \alpha + x \cdot \sinh \alpha$$

Thus

$$\begin{pmatrix} x' \\ y' \end{pmatrix} = \begin{pmatrix} \cosh \alpha & \sinh \alpha \\ \cosh \alpha & \sinh \alpha \end{pmatrix} \cdot \begin{pmatrix} x \\ y \end{pmatrix} = \cosh \alpha \cdot \begin{pmatrix} 1 & \tanh \alpha \\ 1 & \tanh \alpha \end{pmatrix} \cdot \begin{pmatrix} x \\ y \end{pmatrix}$$

$$= \frac{\sqrt{(\cos \alpha)^2}}{\sqrt{(\cosh \alpha)^2 - (\sinh \alpha)^2}} \cdot \begin{pmatrix} 1 & \tan \alpha \\ 1 & \tan \alpha \end{pmatrix} \cdot \begin{pmatrix} x \\ y \end{pmatrix}$$

$$= \frac{1}{\sqrt{1 - (\tan \alpha)^2}} \cdot \begin{pmatrix} 1 & \tanh \alpha \\ 1 & \tanh \alpha \end{pmatrix} \cdot \begin{pmatrix} x \\ y \end{pmatrix} \qquad \text{(Equation 12-6)}$$

For each iteration, if you intentionally set $\tanh \alpha$ to the power of 2, the multiplication in Equation 12-6 will turn into a shift operation. The part $\frac{1}{\sqrt{1 - (\tanh \alpha)^2}}$ can be pre-calculated as constants and stored in a look-up table.

One thing that's unique about hyperbolic coordinate system is its shift sequence. According to Ref [5], some of the shift operations in Equation 12-6 need to be repeated to guarantee convergence. If the shift operation is in the form of $\pm 2^{-i}$, to guarantee convergence, i should start at 1, and the following index should be repeated one more time:

$$i = 4, 13, 40, 121 \ldots k, 3k + 1 \ldots$$

The pre-calculated constant should also take this repetition into account.

Hyperbolic Vectoring Mode

As indicated in Table 12-2, the vector mode in the hyperbolic coordinate system will give you $\sqrt{x^2 - y^2}$ and *arctanh*. The fixed point C model for this mode is presented in Listing 12-19 (theta[i] can be pre-calculated and stored as a look-up table).

Listing 12-19. Vectoring Mode in Hyperbolic Coordinate System

```
void CORDIC_hyperbolic_vectoring_mode (int n,
                          long long *px0, long long *py0, long long *pz0)
{
  unsigned int theta[CORDIC_MAX_ITERATIONS + 1];
  int i,j, iter;
  int sigma;
  int again;
```

```
long long x, y, z;
long long x_tmp, y_tmp;
int repeat_point = 4;

double k = 1;
iter = 1;

x = *px0; y = *py0; z = *pz0;

for (i = 1; i <= n; ++i) {

    if (i == repeat_point) {
        again = 1;
        repeat_point = repeat_point * 3 + 1;
    } else {
        again = 0;
    }

    for (j = 0; j <= again; ++j) {
        if (y >= 0) {
            sigma = -1;
        } else {
            sigma = 1;
        }

        x_tmp = x + sigma * (((y >> (i-1)) + 1) >> 1);
        y_tmp = (((x >> (i-1)) + 1) >> 1) * sigma + y;
        x = x_tmp;
        y = y_tmp;

        theta[i] = floor(((log((1 + (double)1/(1 << i)) / (1 - (double)1/(1 << i)) ) / 2)
        * CORDIC_SCALING_FACTOR + 0.5));

        z = z - (long long)theta[i] * sigma;

        k = k / sqrt ( 1 - 1.0 / ((long long)1 << (2 * i)));
        printf ("--------------------- iter = %d, k = %f\n", iter++, k);

    } // End of for loop

} // End of for loop

*px0 = x;
*py0 = 0;
*pz0 = z;
}
```

For demonstration purposes, the pre-calculated number

$$K(n) = \prod \frac{1}{\sqrt{1+2^{-2i}}} \, (i = 1,2,3,4,4,5,\ldots,12,13,13,14,15\ldots39,40,40,41,\ldots)$$

is crunched in each iteration in Table 12-4.

Table 12-4. Pre-Calculated Constant for Hyperbolic Coordinate Systems

Number of Iterations n	K (n)
1	1.154701
2	1.192570
3	1.201997
4	1.204352
5	1.206711
6	1.207301
7	1.207448
8	1.207485
9	1.207494
10	1.207496
11	1.207497
12	1.207497
13	1.207497
...	1.207497

As you can see, the pre-calculated constant starts to converge on 1.207497 when the n becomes big, and that is the value K_{-1} (here the subscript '-1 stands for m=-1) referenced in Table 12-2.

If you set $x = w + 1/4$, $y = w - 1/4$, $\sqrt{x^2 - y^2}$ can be used to calculate the square root function, as illustrated in Listing 12-20.

Listing 12-20. CORDIC Square Root

```
long long CORDIC_sqrt (long long w)
{
  long long quarter = (long long)CORDIC_SCALING_FACTOR / 4;

  long long x = w + quarter;
  long long y = w - quarter;
  long long z = 0;
  long long result;

  long long Km1 = 1.207497 * CORDIC_SCALING_FACTOR_SQRT;

  CORDIC_hyperbolic_vectoring_mode (CORDIC_MAX_ITERATIONS, &x, &y, &z);
```

```
result = (long long)((Km1 * x + (CORDIC_SCALING_FACTOR / 2))/
                     CORDIC_SCALING_FACTOR);

return result;
}
```

On the other hand, z in Listing 12-19 will converge on $arctanh\left(\dfrac{y}{x}\right)$ according to Table 12-2. Since $arctanh(\delta) = \dfrac{1}{2} \cdot \ln\left(\dfrac{1+\delta}{1-\delta}\right)$, if you intentionally set $x = w+1$ and $y = w-1$,

$$arctanh\left(\frac{y}{x}\right) = \frac{1}{2} \cdot \ln\left(\frac{1+\dfrac{w-1}{w+1}}{1-\dfrac{w-1}{w+1}}\right) = \frac{1}{2} \cdot \ln(w)$$

You can calculate the natural logarithm as shown in Listing 12-21. (The input w should be kept less than CORDIC_SCALING_FACTOR in Listing 12-21 for convergence.)

Listing 12-21. CORDIC Natural Logarithm

```
long long CORDIC_ln (long long w)
{
  long long one = (long long)CORDIC_SCALING_FACTOR;

  long long x = w + one;
  long long y = w - one;
  long long z = 0;
  long long result;

  CORDIC_hyperbolic_vectoring_mode (CORDIC_MAX_ITERATIONS, &x, &y, &z);
  z *= 2;

  return z;
}

//============================================================
// example:
// z0 = CORDIC_ln (CORDIC_SCALING_FACTOR / 2);
// z0 = -744261114,
// z0 / CORDIC_SCALING_FACTOR = -0.693147176876664
// ln(2) = 0.693147180559945
```

Hyperbolic Rotation Mode

As indicated in Table 12-2, the rotation mode in a hyperbolic coordinate system will give you a combination of *sinh* and *cosh*. And the fixed point C model for this mode is shown in Listing 12-22.

Listing 12-22. Rotation Mode in Hyperbolic Coordinate System

```
void CORDIC_hyperbolic_rotation_mode (int n, long long *pxo, long long *pyo, long long *pzo)
{
  unsigned int theta[CORDIC_MAX_ITERATIONS + 1];
  int i,j, iter;
  int sigma;
  int again;

  long long x, y, z;
  long long x_tmp, y_tmp;
  int repeat_point = 4;

  x = *pxo; y = *pyo; z = *pzo;
  for (i = 1; i <= n; ++i) {

      if (i == repeat_point) {
            again = 1;
            repeat_point = repeat_point * 3 + 1;
      } else {
            again = 0;
      }

      for (j = 0; j <= again; ++j) {
          if  (z >= 0)  {
                sigma = 1;
          } else {
                sigma = -1;
          }

          x_tmp = x + sigma * (((y >> (i-1)) + 1) >> 1);
          y_tmp = (((x >> (i-1)) + 1) >> 1) * sigma + y;
          x = x_tmp;
          y = y_tmp;

          theta[i] = floor(((log((1 + (double)1/(1 << i)) / (1 - (double)1/(1 << i)) ) / 2)
          * CORDIC_SCALING_FACTOR + 0.5));
          z = z - (long long)theta[i] * sigma;
      } // End of for loop

  } // End of for loop

  *pxo = x;
  *pyo = 0;
  *pzo = z;
}
```

Since

$$e^{\theta} = \cosh \theta + \sinh \theta$$

If you set both x and y as 1, you can make a natural exponential function out of Table 12-2, as illustrated in Listing 12-23. (The input w should be kept less than CORDIC_SCALING_FACTOR in Listing 12-23 for convergence.)

Listing 12-23. CORDIC Natural Exponential Function

```
long long CORDIC_exp (long long w)
{
  long long one = (long long)CORDIC_SCALING_FACTOR;

  long long x = one;
  long long y = one;
  long long z = w;
  long long result;

  long long Km1 = 1.207497 * CORDIC_SCALING_FACTOR;

  CORDIC_hyperbolic_rotation_mode (CORDIC_MAX_ITERATIONS, &x, &y, &z);

  result = (long long)((Km1 * x + (CORDIC_SCALING_FACTOR / 2))/ CORDIC_SCALING_FACTOR);

  return result;
}

//==================================================================
// Example:
// z0 = CORDIC_exp (562332068)
// z0 = 1812780898
// z0 / CORDIC_SCALING_FACTOR = 1.68828377313912
//
// exp (562332068 / CORDIC_SCALING_FACTOR) = 1.68828386644187
//
```

Summary

This chapter started out with an introduction to the Q format and two's complement numbers.

Basic fixed point operations such as rounding, truncation, saturation were also discussed.

In order to do more sophisticated computations, a full coverage of the CORDIC algorithm was presented, which included three coordinate systems and two operating modes.

References

1. *Digital Signal Processing with Field Programmable Gate Arrays (2nd Edition).* Uwe Meyer-Baese, Springer, 2004

2. "The IBM System/360 Model 91: Floating-point Execution Unit." S. F. Anderson, J. G. Earle, R. E. Goldschmidt, and D. M. Powers, *IBM Journal,* January, 1967

3. "The birth of CORDIC." Jack E. Volder, *Journal of VLSI Signal Processing* 25, 101–105, 2000

4. "The CORDIC Trigonometric Computing Technique." Jack E. Volder, The Institute of Radio Engineers, Inc., 1959

5. "A unified algorithm for elementary functions." J. S. Walther, Hewlett-Packard Company, Spring Joint Computer Conference, 1971

6. *Computer Arithmetic Algorithms 2nd Edition.* Israel Koren, A K Peters, Ltd., 2002

7. "High Speed Arithmetic in a Parallel Device." J. Cocke and D. W. Sweeney, IBM Internal Paper, February, 11, 1957

8. "High-Speed Arithmetic in Binary Computers." O. L. MACSORLEY, Proceedings of the IRE, January, 1961

9. "A New Class of Digital Division Method." James E. Robertson, University of Illinois, Graduate College, Digital Computer Laboratory, Report No. 82, March 5, 1958

10. "Techniques of Multiplication and Division for Automatic Binary Computers." K. D. Tocher, *Q J Mechanics Appl Math,* 11(3): 364-384, 1958

11. *Computer Arithmetic, Algorithms and Hardware Designs.* Behrooz Parhami, Department of Electrical and Computer Engineering, University of California, Santa Barbara, Oxford University Press, 2000

12. "The Mathematics of the Pentium Division Bug." Alan Edelman, Society for Industrial and Applied Mathematics Vol. 39, No. 1, pp. 54–67, March, 1997

CHAPTER 13

■ ■ ■

Popular Ways of Console Communications: Prepare the Tools

If all you have is a hammer, everything looks like a nail.

—Law of the Instrument

Embedded systems is a pandemic topic that concerns many engineering disciplines. Accordingly, a wide variety of tools are used on daily basis by engineers down in the trenches. Although this book has no intention to devolve into a tool shed, it is nevertheless recommended for readers to get familiar with the following tools.

Software Tools

I don't think there will ever be a complete list of software tools, as the software industry is constantly reinventing itself. But if I have to pick, the following are a few ones that come handy for embedded system developers.

Cygwin

If you were an embedded system developer, chances are that you might have to work under both Windows and Unix (or Linux) environments. Cygwin is a handy toolset that allows users to run Unix/Linux commands on a Windows platform. Cygwin is free of charge and can be installed from http://www.cygwin.com.

Text Editor

Choosing an editor is more like choosing your favorite tennis racket. However, good text editors share common traits:

- Support syntax color
- Recognize text files in both Unix and DOS format (I will elaborate shortly)
- Allow column editing and binary editing

© Changyi Gu 2016
C. Gu, *Building Embedded Systems*, DOI 10.1007/978-1-4842-1919-5_13

- Allow you to adjust the indent space and let you use space character in place for tabs

- Allow you to replace the tabs with space or vice versa

- Allow you to do static syntax parsing on languages such as C and jump between function bodies smoothly

Based on these criteria, my personal favorites are UltraEdit (Ref [1]) and Source Insight (Ref [2]). UltraEdit fits all these standards except the last one. And Source Insight is very good at parsing C/C++ languages and cross-referencing between various syntax entities.

If you are working on a shoestring budget and want a free editor, I recommend Notepad++ (Ref [15]) and Eclipse (Ref [3]). One thing great about them is that they both have plenty of third-party plugins thanks to their open source nature. However, beware that although Eclipse is free, not all of its plugins are free!

One nuisance about the text files is that the carriage return (EOL) is a little bit different between the Windows and Unix formats. Windows uses two characters (CR('\r', 0x0D) + LF ('\n', 0x0A)) as a carriage return while Unix only uses one (LF ('\n', 0x0A)). Fortunately, there are tools under the Unix environment that can convert these two formats back and forth, which are called dos2unix and unix2dos, and the same tools are available in Cygwin. In addition, if you choose Subversion to be your source control tool, you can set up the svn:eol-style property to let Subversion automatically do the conversion for you (Ref [8]).

File/Folder Comparison

Engineering is a subtle job. A small change could easily make or break the whole system. It would be handy to have a nice file-comparison tool. This tool would also be helpful when you merge the code of different version. I recommend UltraCompare (Ref [5]) or Araxis Merge (Ref [6]) in this regard. Both are great tools for this job.

Version Control

There are tons of version control tools in this world. A good version control tool should be capable of doing the following:

- Support both command line and GUI access

- Support both Windows (NTFS, FAT32) and Unix (ext2) file systems

- Be easy to use

- Check out files to a local file system

Here I want to spend a few words on the last one. A typical development setting for embedded systems could be one of the two following:

- Everything, including the compiling tool and debug tool, is on Windows platform.

- A hybrid setting, where part of the tools, like ICE/debug tools, are on Windows while other tools, like the cross-compiler, are on Unix/Linux.

If you belong to the second case, I suggest you mounting your Unix home directory to Windows through Samba (Ref [7]) and edit your files on Windows instead of using the arcane vi or emacs. Your version control tool should be able to let you check out files and dump them to a local folder specified by you. In this way, you can focus on your local files most of the time and sync with the server only when you need to.

However, some version control tools (such as clearcase) have a different concept. Take clearcase for example. It implements its own file system called MVFS. You have to mount to its MVFS file system to access all the source files. Each engineer has his or her own view of the source database. This is an innovative idea. I have to give it credit for that. However, such setting brings upon two problems:

- If you are working on your own view and one of the files is modified and checked-in by your co-workers without notifying you, unintended changes that you are unaware of could sneak into your current source project. (You can change your configuration specification to prevent this. There are some tedious steps you have to go through.)

- If on the same project, two engineer groups work at different sites geographically, one of the engineer group might have to suffer long delays when they edit or compile the source code, because all the file R/W have to go to a server that is far away. (Again, you can create some pipe to alleviate this, but only to a certain extent.)

That's why I prefer to have a version control tool checking-out files to local file systems. You only need to access the server when you sync your local files with the server. Multiple sites would not be a serious problem in this case.

Hereby I recommend Subversion (Ref [8][9]) or Git (Ref [16]) to be the version control tool. They are both very powerful tools, but they take drastically different approaches to get the job done.

Subversion (SVN) takes a centralized approach. Only the SVN server has the whole picture (history log, all revisions in the past, etc.), and you have to connect to the server in order to commit changes. On the client side, you can type svn under Cygwin or Linux to access the SVN server through a command line. And Subversion has a nice GUI called TortoiseSVN (Ref [10]) that is smoothly integrated into Windows Explorer.

On the other hand, Git takes a decentralized approach. Each Git clone at a local drive is a standalone database that has everything in it. All the changes are committed to the local database as well. You can push a branch to a remote computer to share your changes with the rest of the world. Git's distributed nature is very suitable for a team that scattered around multiple sites and doesn't have constant access to a central server. Like Subversion, it also has a Windows Explorer GUI extension called TortoiseGit (Ref [17]). In fact, if you are using Git, mostly like you are using the one powered by GitHub (Ref [20]). And GitHub has its own Windows applications to do most things through a GUI.

When it comes to SVN versus Git, I don't think there is one size that fits all. And I will let you readers make your own decisions in this regard.

Issue (Bug) Tracking

To err is human. Even the slightest misstep could throw the system out of whack in a big way. The project team should follow a standard process to keep track of each issue or bug. I am sure those MBAs/PMs would do a better job than I to explain how important a process can be. But if you have a small team and a thin budget to begin with, I recommend the following tools to replace a full-time MBA/PM. :-)

- *Bugzilla*: Bugzilla (Ref [11]) is a lightweight bug tracking tool that originated from mozilla.org. It is written in Perl (starting from V 2.0). If all you need is to track the status of a bug or filing a new ticket, this might be the tool for you.

- *Trac*: Like Bugzilla, Trac (Ref [19]) can also file and track tickets. But in addition, Trac provides its own wiki system, and it can also be cross-referenced to version control systems like Subversion or Git for change-set and history log. So if you want to make a center page for your project with various topics and issues, you might consider using Trac.

- *GitHub*: Although GitHub is known for code hosting with Git, its collaborative workflow makes it suitable for issue tracking, code review, and lightweight project management. More details can be found in Ref [20]. In fact, the companion material of this book is also hosted on GitHub.

But no matter what tools you choose, have a process in place and follow it vigorously. That's what sets an engineer apart from a street artist.

Terminal Emulator

If you are a die-hard fan of the TV show "Lost" (Ref [12]), just like me, you probably would never forget its signature scene. Deep down the dark hatch, a lone man is keying in a mysterious sequence of numbers through a bulky console, and the world was saved once again when he hit the Enter key. :-)

At the time when computer's were invented, it was a big deal to get In/Out through a dummy terminal (console). Despite the progress of technology, the idea of using console for In/Out never dies.

For embedded systems, debug console is an indispensable tool at the early stage of development. It is crucial to be able to get messages to print out and to key in commands to control the target system (DUT). As shown in Figure 13-1, the three most popular ways of console communications are:

- Using JTAG

- Through a 10/100 Ethernet port

- Through the good old RS-232 port

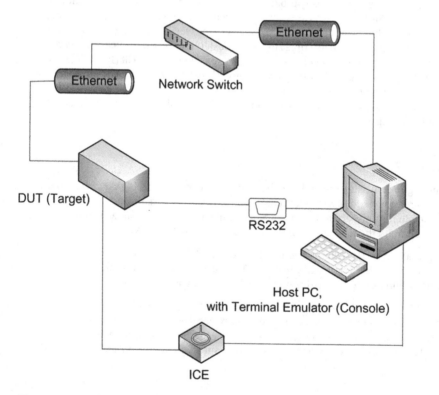

Figure 13-1. *In/out between target and console*

Of course, you don't have to own a VT100 terminal to talk to your target any more. All you have to do is to install a piece of software called a *terminator emulator* on the host PC to mimic the console hardware.

If you use JTAG for your console, you also need an ICE box to go with that. And the ICE software is usually vendor unique.

If your target communicates with your console through Ethernet, you can install either Cygwin or Putty [13]. Your target system should have a server up running with TCP port 22 (ssh) or 23 (telnet) open. This usually means you are able to boot into an embedded OS, or your bootloader is already working properly.

If you are at a cruder stage of your development, by which you have to work in the dark to figure out the code flow, RS-232 is probably the only viable way, both economically and technically, to fulfill the debug console function. And here I strongly recommend Tera Term (Ref [14]) to be your serial console software. It has very flexible scripting capability and is the best terminal emulator I have ever seen. Its latest version also supports the ssh protocol.

Virtual Machine

Today's desktop PCs are so powerful that running a CPU simulator hardly sweats them. Among all the virtual machine platforms, two stand out:

- Virtual Box (`https://www.virtualbox.org/`)
- VMware (`http://www.vmware.com/`)

For embedded system development, virtual machine can be used to host Linux systems. There are cases where Linux platform is required to run tool chains, and virtual machine can be a low-cost alternative if you have a small budge and want to save cost on hardware.

Cluster Workload Management

For those who are doing heavy load computations, like RTL simulation or regression tests of any kind, having a cluster of computers to divide and conquer the work load will become necessary. For that purpose, you need good workload management software, such as Platform LSF (Ref [28]) or Openlava (Ref [29]).

In fact, those two bear striking resemblance to each other, as the latter is an open source fork of the former's early version (version 4.2 probably). And you could also let those clustering software manage your floating licenses so that jobs can be scheduled more efficiently. For overnight tasks, e-mail notification is also available.

Hardware Tools

If you were a soldier in the battlefield, you probably would wish to have all the cannons and tanks on your side. For engineers, here is a list of your cannons and tanks.

Oscilloscope

During the project review stage, don't forget to ask board designers to leave enough test points or ground leads for signal probing purposes. Also make sure you have the right probe and scope for the frequency you are dealing with and calibrate them before using.

A good introduction to oscilloscope basics can be found in Ref [21]. Most oscilloscopes allow you to change its internal termination between 50 Ω and 1MΩ (or maybe 10 MΩ). For most of the time, you should set the termination to high impedance to avoid disturbing the circuit under test. However, for some RF circuit, you might want to set the impedance to 50 Ω to observe the waveform under a loaded condition.

Logic Analyzer

First of all, if there is a FPGA in your system and that FPGA happens to have enough memory left unused, you can turn the FPGA into a small scale logic analyzer. That saves you a lot of debugging costs.

If this is not a viable option, you have to do the following to hook up a logic analyzer for real:

1. During the board design stage, make sure you have space left for probe connectors. In the old days when clock rate is low, a 40-pin single end pod connector might suffice. But nowadays the connectors all have very small pitch and can be in various forms to reduce parasitic inductance. Ref [22] shows a lot of possible form factors.

2. Get a probe to match the connectors on board. Depending on the probing solution you choose, the probe/cable could cost you more than a thousand bucks.

3. Get a logic analyzer that has enough memory. The setup of logic analyzer usually involves timing/state mode, triggers, etc. And a good introduction can be found in Ref [23].

Keep in mind that all these steps have dollar signs involved!

Bus/Protocol Analyzer-Exerciser

Depending on how complicated the bus/protocol you are dealing with, a bus/protocol analyzer (or Exerciser) might be indispensable. Ref [24][25] are example ones for I2C/SPI and PCI bus, respectively.

Power Analyzer

Power consumption is an important metric for embedded systems. One way to measure power is to measure the current on power rail with two-terminal sensing, as shown in Figure 13-2. A sensing resistor in the order of mΩ is put in serial with the power rail. The voltage across it is measured by a voltage meter.

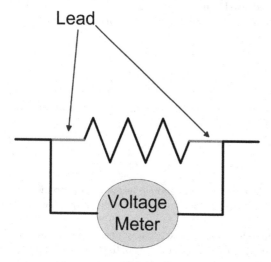

Figure 13-2. *Two terminal sensing*

The biggest problem with two-terminal sensing is its accuracy. In two-terminal sensing, the current value is derived per Ohm's law $(I = V / R)$ through the voltage reading and the normal value of the sensing resistor. However, due to the contact resistance or lead resistance, the actual resistance will inevitably deviate from its normal value. The temperature fluctuation on the lead will also contribute to resistance variation.

As suggested in Ref [26], one way to improve the accuracy is to replace the two-terminal resistor in Figure 13-2 with a four-terminal one. In this way, the lead length can be reduced to a minimum.

And the idea of four-terminal setting also leads to a different measurement method called Kelvin[1] configuration, as shown Figure 13-3. The Kelvin configuration reads out both voltage and current to accurately measure load power consumption, and it is often the method adopted by Power Analyzers (Ref [27]). Power Analyzer can integrate over a long period of time to average out the fluctuation, and it is also indispensable to measure standby power. I suggest getting one if your budget allows.

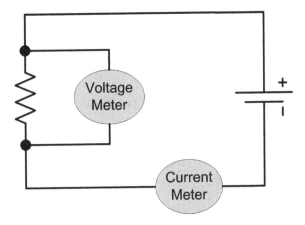

Figure 13-3. *Kelvin configuration*

Multimeter

Among all the hardware tools mentioned in this chapter, Multimeter is probably the least expensive. So get one for obvious reasons. :-)

Others

Depending on the particular domain you are working with, there might be tons of domain specific ones you have to spend money on. (If you want to get a feel of that big dollar sign, ask an RF engineer to see how many pieces he needs to begin with.)

Summary

This chapter provided a brief introduction to the tools for embedded system development, on both the hardware and software sides. The list presented here is by no means complete. But if you are looking for advice to fill your capital spending, this is the shopping list to begin with.

---

[1]Named after William Thomson, Lord Kelvin

References

1. UltraEdit, IDM Computer Solutions, Inc. (http://www.ultraedit.com)
2. Source Insight, Source Dynamics, Inc. (http://www.sourceinsight.com)
3. Eclipse Foundation (http://www.eclipse.org)
4. IDM Extra Downloads, IDM Computer Solutions, Inc. (http://www.ultraedit.com/downloads/extras.html)
5. UltraCompare, IDM Computer Solutions, Inc. (http://www.ultraedit.com/products/ultracompare.html)
6. Araxis Merge, Araxis Ltd. (http://www.araxis.com/merge/index.html)
7. Samba (http://www.samba.org)
8. Version Control with Subversion: For Subversion 1.8: (Compiled from r5039). Ben Collins-Sussman, Brian W. Fitzpatrick, and C. Michael Pilato, 2013
9. Apache Subversion (http://subversion.apache.org)
10. TortoiseSVN (http://tortoisesvn.net)
11. Bugzilla, (http://www.bugzilla.org)
12. "Lost" (TV Series), ABC.com (http://abc.go.com/shows/lost)
13. PuTTY: A Free Telnet/SSH Client (http://www.chiark.greenend.org.uk/~sgtatham/putty/)
14. Tera Term Open Source Project (http://ttssh2.osdn.jp/)
15. Notepad++ (https://notepad-plus-plus.org/)
16. *ProGit—Everything you need to know about GIT (2nd Edition)*. Scott Chacon and Ben Straub, Apress, December, 2014
17. Tortoise Git (http://tortoisegit.org/)
18. Bugzilla Documentation, Release 5.0rc3+, The Bugzilla Team, May 27, 2015
19. Trac, Integrated SCM & Project Management (http://trac.edgewall.org/)
20. GitHub (https://github.com/)
21. XYZs of Oscilloscopes, Tektronix, 2011
22. Probing Solutions for Logic Analyzers, Agilent Technologies, 2011
23. The XYZs of Logic Analyzer, Tektronix, 2001
24. Aardvark I2C/SPI Host Adapters User Manual V5.15, Total Phase, Inc. Feb 28, 2014
25. TA700/800 Series PCI-X, PCI, CPCI, PMC BUS Analyzer-Exerciser User's Manual, Catalyst Enterprises, Inc., 2005
26. 2-Terminal vs. 4-Terminal Resistors, RCD Application Note R-31, RCD Components Inc.
27. PA1000 Power Analyzer User Manual, Tektronix
28. IBM Platform LSF, IBM Technical Computing Data Sheet, IBM Corporation, IBM Systems and Technology, October, 2013
29. OpenLava, Open Source Workload Management (http://www.openlava.org)

CHAPTER 14

■ ■ ■

Work Flow

> One way to identify an amateur organization of any sort, be they accountants, lawyers, craft shops, or software developers, is a lack of process.

> —Jack Ganssle, Embedded Systems Design (September, 2003)

There is more than one way to design an embedded system, and there are more than 100 ways to design a system that has defects. For any engineering work, having a process in place is a sound way to ensure consistency and reliability.

However, process is often deeply tied to the business nature of the company, and probably no two are exactly the same. For embedded system design, this chapter only serves to paint a general picture, as demonstrated in Figure 14-1. Feel free to tailor the flow below to fit your actual circumstances.

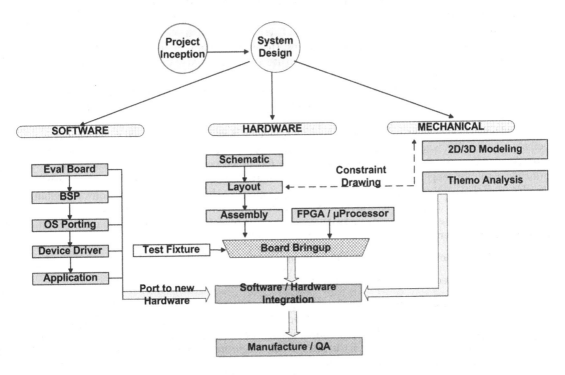

Figure 14-1. General work flow

© Changyi Gu 2016
C. Gu, *Building Embedded Systems*, DOI 10.1007/978-1-4842-1919-5_14

Project Inception

The seed of a project can come from multiple sources. It could be the brainchild of your product managers, or it could come as a reward by winning bid. The inception of a project often comes with a budget and a schedule, based on which the project manager and project engineer will start to do the planning and system partitioning.

To increase the chance of success, trade study and cost analysis are absolutely necessary at this stage to show that a viable product can be made in a profitable way. And spending enough hours at the planning stage is always a wise move in my opinion.

System Design

The project lead will call on engineers across all disciplines for their opinions, and make decisions at system level that could fit the given budget and schedule. For embedded systems, the following three groups often stand out: Software, Hardware, and Mechanical.

Software

The type of processor has to be determined at the system design stage. While waiting for hardware engineers to deliver the board, the software team can start by working on an eval board that has the same type of processors. Hopefully, the final hardware will be a close sibling of the eval board in use, with additional peripherals or modified memory mapping.

Most processor vendors today will go the extra miles to provide a full suite of BSP (Board Supporting Package) along with the bare hardware. The supporting package often includes the bootloader (such as U-Boot, RedBoot, etc.), a HAL (Hardware Abstraction Layer) or embedded OS (Linux, VxWorks, etc.), drivers for the peripherals, and the test/demo utilities. The software team can use them as a base for new development. If feasible, the software team could also use a daughter card to develop drivers for those additional peripherals soon to be introduced in the final product.

Depending on the circumstances, application software can even be developed on a simulated platform without any physical hardware involved, so that multiple groups can work in parallel to meet the targeted schedule.

And I believe every software engineer has come across the "Software Engineering" subject at some point in his or her life, whether it is in school or at work. And inside every software team, there should be more detailed flow to enforce consistency on code style, tool usage, version control, etc. to stick to the principle of software engineering. However, the phrase "Software Engineering" is a non-linear time-variant variable (Ref [1][2] :-). New flows are constantly being invented as to how to make reliable software within the shortest period of time. Chapter 6 includes as a generic flow that is used by many. And inquisitive readers can Google the following to witness the pandemonium of "software engineering" and scoop useful information out of them:

- XP (Extreme Programming), Agile, Scrum

- The Cathedral and the Bazaar (Ref [3])

Among all the hypes and hopes offered here, I found the CI (Continuous Integration) concept to be relevant to the embedded system development, especially if you run software simulation quite a lot. To put it succinctly, CI advocates frequently building and testing for each commit (or nightly build/test) to the trunk (for subversion) or master branch (for Git). In this way, the trunk/master branch can be kept in a deployable state most of the time to avoid "integration hell".

Practically, it usually means you have to set up a CI server, such as Jenkins (Ref [4]), and do build and unit tests after every commit made to the version control system. For embedded systems, the unit test might be a little bit hard to carry out if extra hardware has to be set up and be constantly available. However, for the case of software simulation, to build/test as much as you can is a viable approach to eliminate bugs at the early stage.

Hardware

While software engineers are working wee hours, their hardware counterparts are burning midnight oils as well.

The PCB design starts with a schematic, which is heavily tied to the BOM cost. And the eval board with the intended processor can serve as a good reference. A joint review should be held before the schematic is sent to layout. My two cents is that the review should also invite firmware engineers in order to best accommodate their debug wishes.

In addition, if FPGA is involved, a preliminary test build should be conducted by FPGA engineers to make sure the pin assignment is feasible. Sometimes, especially for IPs like DDR controller or high-speed transceivers, pins have to be assigned to certain IO banks or differential pairs for IPs to fit. A preliminary test build can rule out possible surprises down the road.

The PCB layout can be done either in house or contracted out. Layout engineers often need to work with various parties like the schematic designer, the mechanical engineers, or even the product manager to determine the constraint drawing, the component facing and placement, etc. The number of layers also needs to be determined to find a good comprise between signal integrity and cost.

If there is FPGA or CPLD on the board, the layout engineers can request FPGA engineers and schematic designers modify the pin assignment to make the placement and routing less painful.

Depending on the turn-around time and cost, the bare board and assembly work can be done locally or overseas. When the assembled boards come back, the hardware engineers should do a basic smoke check to make sure there is no shorting, and start powering on the components section by section to gradually bring up the board. Firmware engineers could get involved at any stage for testing and integration.

To facilitate debug, many PCBs with a small form factor would have a test fixture or daughter card designed separately to help probe the board.

If FPGA is in the design, the FPGA engineers should work in parallel to simulate the design as much as they can and deliver bit stream files for integration. Since these days the FPGAs and firmware are so intertangled, the simulation could include the processor behavior models as well without any hardware physically available.

Mechanical

Mechanical engineers can make the 2D/3D model before the hardware is built for product manager and project lead to review. Thermo analysis is also part of their jobs. This book will leave the part of mechanical design to inquisitive readers.

Manufacturing and Quality Control

Process is important in the design stage, and it is more so for manufacturing. The design engineers should have concepts like DFT (Design for Test), DFM (Design for Manufacture), etc., in mind if the final product is intended for mass production. And before mass production, the product should go through multiple verification phases, like EVT (Engineering Verification Test) and DVT (Design Verification Test). Depending on the volume and cost, the manufacturing can be done in house or through a contracted manufacturer. Again, I will leave the details to inquisitive readers.

Summary

McDonald's is able to grow itself into nearly every corner of the earth because it has a standard process that each and every store follows. For engineers, a standard process is what it takes to deliver a dependable product consistently.

The process mentioned in this chapter only serves as a reference model. You can adopt, modify, or come up with your own process as long as it meets your business needs. But make sure you follow your own principles once they are in place.

References

1. "Is Software Engineering an Oxymoron?" Allen Holub, March, 15, 2005, *SD Times*
2. "Software Engineering Is NOT an Oxymoron," Jack Ganssle, March 23, 2005, Embedded.com
3. The Cathedral and the Bazaar, Version 3.0. Eric Steven Raymond, Copyright 2000, (http://www.catb.org/~esr/writings/cathedral-bazaar/cathedral-bazaar/)
4. Jenkins, an extensible open source continuous integration server (http://jenkins-ci.org/)

Index

Get the eBook for only $5!

Why limit yourself?

Now you can take the weightless companion with you wherever you go and access your content on your PC, phone, tablet, or reader.

Since you've purchased this print book, we're happy to offer you the eBook in all 3 formats for just $5.

Convenient and fully searchable, the PDF version enables you to easily find and copy code—or perform examples by quickly toggling between instructions and applications. The MOBI format is ideal for your Kindle, while the ePUB can be utilized on a variety of mobile devices.

To learn more, go to www.apress.com/companion or contact support@apress.com.

Printed in the United States
By Bookmasters